Laser Safety

Practical knowledge and solutions

IOP Series in Coherent Sources and Applications

About the Editor
F J Duarte is a laser physicist based in Western New York, USA. He has a 30+ year experience in the academic, industrial and defense sectors. Duarte is editor/author of 13 laser optics books and sole author of two books (*Tunable Laser Optics* and *Quantum Optics for Engineers*). He has made original contributions to the field of narrow-linewidth tunable laser oscillators, organic laser gain media, nanoparticle solid-state laser materials, and laser interferometry. He is also the author of the multiple-prism grating dispersion theory applicable to tunable lasers, laser pulse compression, and coherent microscopy. Duarte is a Fellow of the Australian Institute of Physics (1987) and a Fellow of the Optical Society of America (1993). He has been awarded the Paul F Foreman Engineering Excellence Award and the David Richardson Medal from the Optical Society.

Coherent Sources and Applications
Since the discovery of the laser, applications of this wondrous emitter of coherent radiation have grown enormously. Subsequently, we have also become familiar with additional sources of coherent radiation such as the free electron laser, optical parametric oscillators, and interferometric emitters. The aim of this new book series is to explore and explain the physics and technology of widely applied sources of coherent radiation and to match them with utilitarian and cutting-edge scientific applications. Selected coherent sources are those that offer advantages in particular emission characteristics areas such as broad tunability, high spectral coherence, high energy, or high power. An additional area of inclusion is those coherent sources capable of high performance in the miniaturized realm. Selected uses include practical applications valuable to the industrial, commercial, and medical sectors. Particular attention will be given to scientific applications with a bright future scope such as coherent (or laser) spectroscopy, astronomy, biophotonics, space communications, space interferometry, and quantum entanglement.

Publishing benefits
Authors are encouraged to take advantage of the features made possible by electronic publication to enhance the reader experience through the use of colour, animation and video, and incorporating supplementary files in their work.

Do you have an idea of a book that you'd like to explore?
For further information and details of submitting book proposals, see iopscience. org/books or contact Ashley Gasque at ashley.gasque@iop.org.

Laser Safety

Practical knowledge and solutions

Ken Barat

42179 W. Santa Fe St, Maricopa, Arizona, AZ 85138, USA

IOP Publishing, Bristol, UK

ISBN   978-0-7503-1692-7 (ebook)
ISBN   978-0-7503-1690-3 (print)
ISBN   978-0-7503-1691-0 (mobi)

DOI   10.1088/2053-2563/ab0f25

Version: 20190601

IOP Expanding Physics
ISSN 2053-2563 (online)
ISSN 2054-7315 (print)

British Library Cataloguing-in-Publication Data: A catalogue record for this book is available from the British Library.

Published by IOP Publishing, wholly owned by The Institute of Physics, London

IOP Publishing, Temple Circus, Temple Way, Bristol, BS1 6HG, UK

US Office: IOP Publishing, Inc., 190 North Independence Mall West, Suite 601, Philadelphia, PA 19106, USA

The contributing authors are what make this text stand out from other texts on laser safety. It is to them that I owe the greatest gratitude. I must also include by grand-daughter Margo, who never fails to bring a smile to my face. In today's world that is priceless. Hello Margo, it is Grandpops.

Contents

Part III Not the usual topics, going outside the space–time continuum

Preface

Some technologies come along and seem to capture the world. They are everywhere. The automobile is a good example. They were around for a few years as a novelty and then with the introduction of mass production their use exploded across the world. I am sure the reader can think of some other examples. To me the flat screen replacing the cathode ray tube monitor is another example. Who would think they would grow into billboards and more.

The laser is another example. Thomas Maiman produced the Ruby Rod laser in 1960 and after several years of development of new types of lasers, laser technology has spread to all sectors of society.

The earliest date I am aware of for the first reported laser accident is 1964. The goal of this book is to not only explain and review the basics of laser safety, that is keeping the user, those around them and laser equipment free from harm, but to go over a number of helpful topics.

Many of these topics seem to be so obvious one might wonder why they are here. The answer 'skill of the craft' just like 'common sense' is not as widespread or known as one likes to think.

The contributing authors and myself hope you find this book useful enough to suggest to others that they read it.

Ken Barat

Editor biography

This book has a number of chapters contributed by some of the finest people involved in laser safety and related work. I greatly appreciate all who gave up their time to contribute to this text. I hope you, the reader, will feel the same way.

Ken Barat, CLSO

The former Laser Safety Officer for Lawrence Berkeley Nat Lab and the National Ignition Facility Directorate. Presently, he is providing laser safety consulting under the title of Laser Safety Solutions. He is the author of several texts on laser safety as well as numerous articles and presentations worldwide.

He is an OSA distinguished speaker, Fellow of the Laser Institute of America, and a senior member in IEEE and SPIE. He was the chair and organizer of the first seven LSO workshops. ANSI committee member and chair. For laser questions he is the 'Ask the Expert' for the Health Physics Society. Former LIA board member and laser safety instructor for several laser professional societies and institutions. A Rockwell award winner, and among the first class of Certified Laser Safety Officers.

List of contributors

Patrick Bong

Engineer at Lawrence Berkeley National Laboratory, With 20 years of experience in a research environment, developing interlock strategies for particle accelerators, x-ray light sources and laser systems, my objective is to share my experience and learn from others in widely dissimilar fields who perform similar work. Instructor at US Particle School.

Karen Kelley

Karen Kelley, CIH, CLSO, is the Director of Research Safety at Cornell University. She has a BS in Mathematics, an MS in Environmental Health, and over 20 years of experience in research safety, industrial hygiene, and laser safety in both industry and academia. Karen previously served as the Laser Safety Officer at the University of Pennsylvania and the University of Maryland. A focus of her work over the past few years has been on safety culture in academic research. Karen also contributed a chapter to *Understanding Laser Accidents* 2018.

Tom MacMullin

President and General Manager of Kentek Corporation. A driving force in laser safety for 20 years. Profit and Loss responsibility for manufacturer and distributor of laser accessories including laser safety eyewear, laser curtains and barriers, flash lamps, laser cavities, laser rods, laser power meters, laser beam detection devices and other components. Specialties: technical business leader, innovative strategic thinker, energetic team builder. MBA, MIT Sloan School, Cambridge, MA.

Lisa Manglass

Lisa Manglass is a Health Physicist and PhD Student at Clemson University in the Department of Environmental Engineering and Earth Sciences. She holds a BS in Physics from the University of Georgia and an MS in Health Physics from Colorado State University and worked for seven years as an environmental health physicist prior to entering the PhD program at Clemson in 2016. As a consultant, she specialized in radiological risk assessment and radiological characterization. Lisa is an associate member of the American Academy of Health Physics.

Randy Paura, P. Eng., CLSO

President of Dynamic Laser Solutions Inc., graduate of the University of Waterloo, Ontario, Canada and a Professional Engineer of Ontario since 1991. Randy is an industrial laser materials processing and safety consultant, providing independent, third party laser system solutions, LSO training, laser safety conformance and process audits. He is the Chair for Z136 TSC-2 for laser hazard and classification evaluation, Vice-Chair for Z136 SSC-9 safe use of lasers in manufacturing environments and member of AWS C7C American Welding Society, Subcommittee on Laser Welding.

Michael Thomas

Michael Thomas is President and founder Spica Technologies Inc. Spica was founded in 1992 to provide unique optical test and metrology solutions, and specializes in laser damage measurements, laser metrology and laser based optical density measurements on laser protective devices, including filters and eyewear. He holds a BS Optics from the University of Rochester and is an active member in the ISO/TC-94/SC-6 group developing new laser eyewear standards. Mr Thomas has authored or coauthored over 30 technical papers and holds multiple patents.

Part I

The basics of laser safety

IOP Publishing

Laser Safety
Practical knowledge and solutions
Ken Barat

Chapter 1

Why are laser accidents still happening?

Ken Barat

Why are laser accidents still happening? One answer is that people do not find out about accidents. Most institutions keep the information to themselves. They fear negative press. Rather than making the information open to help prevent similar accidents from occurring somewhere else, they keep it to themselves, like a card player in a poker game. I know of several major academic institutions who have had incidents which prompted major changes at their sites but who will not share the information. They will have to live with the guilt when a similar incident happens elsewhere, or they will not have any guilt because no-one will find out that it happened somewhere else.

So, I ask again, why do laser accidents happen? The answer changes with the laser setting, and changes with the most popular safety philosophy of that year. Two answers that I think will endure the passage of time are, no negative consequences to bad behavior and poor to a lack of mentoring. These two items are the ones I would like to explore in this chapter.

1.1 Bad behavior and no negative consequences

The above title seems to contradict itself. But once one thinks about it, the words take on a clear meaning and highlight a major contributor to not only laser accidents, but all accidents, excluding equipment failure. So, let me clearly state—this chapter deals with human actions.

Statistics show us that most laser accidents occur to experienced laser users. The question of why seems to be a natural one. An article I recently read seems to offer one answer[1].

It started with the question, why do people, including myself, drive over the posted speed limit? Well the answer is the behavior has occurred from hundreds to

---

[1] *The Human Factor in Safety and Operations* by David Mallard, EHS today June 2016, www.EHStoday.com.

thousands of times without negative consequences. If one got a ticket each time you exceeded the speed limit your behavior would change, and change rapidly.

The same applies to experimental laser users. Not checking for stray reflections, peaking under/over eyewear, not wearing eyewear, each time we violate good laser safety practice and get away with it, that action re-enforces the poor practice, I am not advocating for injuries, just stating a fact. If one was hit in the eye each time you looked under one's eyewear, no doubt the practice would stop.

Looking at the list of behaviors below, which are usually credited as the cause of most laser accidents, how many of these get a contributing factor from the fact that users have frequently repeated them before something happens to them, causing them to say 'Wow that was a close call, do not do that again'.

(a) **Unanticipated eye exposure during alignment**

It is a universal law that beam manipulation can produce unexpected reflections. It is this activity that separates R&D laser work from all other laser use settings, where beam manipulation is an exception, or only service activity, not an almost daily activity. Therefore, good alignment practices are critical to preventing laser accidents.

(b) **Misaligned optics and upwardly directed beams**

Unless one is working with vertical breadboards, lasers are thought of in the horizontal plane. Hence, a risk is present any time beams are directed upward to change beam height. So, one needs to be aware of this condition and take steps to protect yourself and others.

(c) **Available laser protective eyewear not used**

Yes, this happens all the time. It all traces back to one of two reasons. The eyewear is a problem, poor visibility, excessive weight, poor fit, etc, a list of reason and excuses for the user not to use their eyewear. The other reason is just being lazy/over confident. The feeling that the user is so familiar with the activity that the eyewear they do not wish to use is not needed.

(d) **Equipment malfunction**

Many times, this happens due to a lack of a preventive maintenance program or consideration of such. Environmental conditions can also add to this, e.g. dirt, blocked ventilation holes, poor handling or storage.

(e) **Improper methods of handling high voltage**

This can be a simple as the lack of grounding or improper grounding, to frayed wires, the list on electrical safety goes on.

(f) **Intentional exposure of unprotected personnel**

Yes, there have been a few such cases. So, follow the golden rule, treat others as you would have them treat you.

(g) **Operators unfamiliar with laser equipment**

This goes back to improper on the job training/mentoring. Sometimes, it is the individual wishing to appear more experienced or knowledgeable to others than they are. In some cases, it is having worked with similar equipment and, therefore, believing they can figure it out or it is all the same. This is very common in industrial settings, counting on 'skill of the craft'.

(h) **Lack of protection for non-beam hazards**

While we focus on laser hazards, these are very commonly not the greatest risk to the laser user. As I have said and written before, laser eyewear does very little if one is electrocuted. The same goes for chemical burns or suffocation from gases.

(i) **Improper restoration of equipment following service**

Whenever service is performed on equipment, there should be a verification that any interlock bypasses have been removed and safety systems are back functioning as expected.

(j) **Laser protective eyewear worn not appropriate for laser in use**

This is most common in laser use areas that have multiple pairs of eyewear for different laser use. Therefore, awareness of what the activity is and checking one's eyewear is critical.

(k) **Unanticipated eye/skin exposure during laser usage**

Most commonly this occurs when putting one's hands in the beam path, an exposure from a stray reflection or a beam escaping a broken fiber

(l) **Inhalation of LGAC and/or viewing laser generated plasmas**

Not using a smoke evacuator or using a poorly functioning one. The worst case would be not using any means to capture the plume due to being unaware of its risk to one's health.

(m) **Fires resulting from the ignition of materials**

Being aware of the irradiance of the beam and the combustion potential of materials being used. A simple example is when I have seen people use post-it notes as beam blocks for stray reflections.

(n) **Eye or skin injury of photochemical origin**

From unblocked UV scatter, which is not felt while one is being over exposed, as well as the effect showing up or being felt hours later, as many as 10–12 h.

(o) **Failure to follow standard operating procedures (SOPs)**

If the SOP is instructional it should be followed, but many times users do not develop a SOP they feel has any value other than to make the Safety Department feel good or to satisfy safety rules.

(p) **Introduction of foreign materials (cables, pages of loose paper, paper clips, falling items or objects)**

We never want unintended items to fall in the beam path, this goes back to housekeeping or set-up procedures

(q) **Modification of the beam path**

There is nothing wrong with modifying beam paths as needed, but one must realize any such action presents the opportunity for stray reflections which can cause injury if not blocked or mitigated.

(r) **Poor communication**

Communication is the key to safety, each person working on the lab must be aware of, or keep aware of, the actions of others. Especially if one person has removed a beam block, moved an optic etc. This extends to entering information in research logs, so people can read what changes may exist before they start their work.

(s) **Turning to look at source of bright light detected by peripheral vision**
Unexpected bright lights or flashes are never a good sign and how one looks at them or reacts to them can lead to more trouble.

1.2 Possible future: virtual reality

While not commonly used there are a few facilities and some software that allow a user to use virtual reality (VR) to explore a laser room, including the feel of performing alignment. It is clear, as VR becomes more accessible, that this approach could become a common training aid for laser users and laser technicians. Keep in mind that the flight simulator had a significant and documented impact on reducing aircraft crashes because the flight crew could experience and practice responses to unanticipated challenges, such as a bird being sucked into an engine or a single engine failure. The VR approach is an excellent bridge to the next section in this chapter.

1.3 On the job training/mentoring (a topic you will find mentioned several times in this text, but learning is through repetition)

Despite efforts to improve pilot performance, crashes due to pilot error remained at 65% for more than 50 years. That changed in 1990 when flight simulators for pilot training became standard, which provided a tool designed to provide experimental learning in a safe and controlled setting. Since then, crashes due to pilot error have declined by more than 54%, now fewer than 3.4 defects per 1 million opportunities.

For us in the laser community, the days of flight simulators are not that far off. I have already encountered VR units to practice laser alignment.

But for most laser users, it is mentoring (on the job training, OJT), being taught good practices and seeing them practiced, that makes the difference. So, they become the expected mode of operation and not the exception. The majority of incidents occur in activities that are perceived to be safe, that is when one is most likely to cheat on the way it should be done. Perceived safety comes from violating good practice with no consequences. If not for our own protection, but for the safety of those around us we cannot accept risk taking during laser work. Laser eye injuries are like a ball in a roulette wheel, one does not know where it will land or how severe the injury might be. Get smart, be safe, I am now stepping down from the soapbox.

The greatest risk of OJT or mentoring is that of mentoring bad habits. Sometimes this is presented to new users as 'tricks of the trade'. Unfortunately, they do not have the experience to understand the basics or risks these so-called tricks/shortcuts present.

Bad habits are not only deliberately learned activities, but more commonly, observed bad habits. When the senior person does not wear laser protective eyewear during alignment or beam manipulation the novice sees this and gets the message that it is the way things are done.

If unexplained, they may miss that a shutter has been put in place, so the risk has been removed or other mitigation is in place, if that is the case.

Laser work in research has elements of art built into it. How the artist handles the brush will yield different effects on the canvas. How one learns to handle optics and mounts will affect the quality of work and the time it takes to set up work. This covers items like finger prints on optics, setting the initial screw settings on optical mounts, labeling mounts, the list goes on.

1.4 Recommendation

OJT needs to be documented. No-one turns on an expensive laboratory system and says 'GO for it'. There is always some orientation period. That period should involve instruction, observations, a question period and finally approval. The trainer and trainee both need to sign a form that they are satisfied the trainee can work on their own. This can be done in a step-by-step process or an all-in-one approach. By documenting this agreement all parties go on record that a certain level of competence has been reached.

1.5 Trainer, what are your responsibilities?

One of the key tasks the OJT instructor must perform when meeting with a new trainee is to review the entire goals of the OJT with the trainee. The OJT instructor should emphasize the expectations of the training, including the skills the trainee will need to acquire to be fully qualified. The instructor will need to explain how the training will be conducted, how the trainee will be able to successfully achieve each task, the limits on the trainee and how they will change over the OJT process.

The trainer needs to explain what he or she is doing, why he or she is doing the tasks, and the safety precautions that must be considered to do the job safely. A common trainer problem is to overwhelm the trainee with details on the first day.

1.6 Trainee, what are your responsibilities?

Ask questions, don't be afraid to look uninformed, so again *ask questions*.

A trainee should ask questions for clarity when in doubt, keep notes and look for additional supporting self-study materials.

The steps of the process need to include the following items:

1. **Demonstrate how to perform a task.** The OJT instructor should select a simple task and demonstrate how to perform the task correctly, explaining why the task is to be performed, its importance, and the impact to the operation of the part or system if not done correctly.
2. **Allow the trainee to perform part of the task.** In this phase of the instruction, the trainer and trainee are interacting together, with the trainer coaching as necessary.
3. **Allow the trainee to perform the entire task.** This should be done with coaching as necessary from the OJT trainer. Depending on the task complexity, this may occur several times before the task is mastered.

4. **Evaluate the trainee's performance of a task.** Observe the trainee performing the entire task without supervision. When the trainee can perform the task without supervision, he or she is considered trained. A trainee OJT record can be constructed so that each series of tasks can be signed off as an accomplishment.

5. **Sign off the training package.** To confirm their competency, the trainee should be allowed to perform the complete task without active supervision. At this point, the training for that particular task will have been completed, and the instructor can sign off.

6. **Keep records.** The final signature page should be filed in a secure place. This may be in the form of physical storage or a digital storage system. The training record should be maintained for as long as the organization deems necessary. For some organizations, the minimum time for maintaining training records is five years.

1.7 What makes good coaching?

Do you praise, curse or just roll your eyes? One of the most difficult jobs the OJT instructor faces is providing trainees with feedback about their progress. Giving the impression that the trainee is progressing well when they need more coaching is counterproductive and dishonest to the trainee. On the other hand, being too brisk or frank about a trainee's progress can easily be mistaken for criticism, especially if there are no established criteria used to evaluate the trainee's performance. Feedback provides trainees with an idea of how well they are performing. Therefore, one should:

- Provide an immediate and complete answer to the task item after the trainee's completion; all parts of the answer or answers should be provided. If alternative methods for completing the tasks are acceptable, each should be included.
- Promptly give the trainee practical means to better understand a task in which he or she underperformed. For example, go over the task and the procedures to which the task relates and then have the trainee practice under your supervision.
- Provide guidance for remediation. The purpose of feedback is to help trainees learn the material. Therefore, OJT should be designed so that the trainee is led to restudy the information he or she failed to recall, recognize, or perform.

1.8 Training review

Once permission has been granted for an individual to work by themselves or unsupervised there is nothing wrong with the instructor coming back after a while to see if the individual is following good technique or has developed some of their own workaround that may not be at the safety level one wants or is acceptable. Hey, they may have found a better way to get the job done and can teach you. Education never ends.

Chapter 2

Classification: a means of hazard communication

Ken Barat

2.1 Introduction

Laser classification, or more formally laser hazard classification, is in simple terms a means of hazard communication. By knowing the class of the laser or laser system one knows its potential to cause harm.

Laser classification is used initially to define the hazard potential of a laser product, this holds true no matter where in the world the laser product is manufactured. Once the class is defined by the manufacturer, classification becomes important to the laser safety officer since the highest classification of the laser in use sets the control measures that need to be considered for safe operation. Please note: whether one needs a laser safety program (and many of its elements) does not depend on the number of lasers one has but on the class of the lasers and the potential for human exposure. One Class 3B or Class 4 laser sets the requirement for a laser safety program. This chapter will not touch on controls, I am saving that for another chapter.

The three major organizations where one finds definitions of laser hazard classes are: Center for Devices and Radiological Health (CDRH) a branch of the United States Food and Drug Administration, American National Standards Institute (ANSI) Z136 Series, Safe Use of Lasers. These two are used in the United States. Then there is the International Electrotechnical Commission (IEC) 60825 series chiefly 60825-1. Which is the primary backbone of classification throughout most of the world. There are slight differences between lasers classified in the United States and the rest of the world. Overall there is a strong desire to harmonize all product classification approaches.

2.2 KEY ITEM about laser hazard classification and why people care

The need for, or better yet, the requirement to have a laser safety program is all based on what classes of lasers your staff may be exposed to. This fact is mentioned in the second paragraph of this chapter, but for those of us who skim over items it is worth repeating. It is not the number of lasers one is around it is the class of the laser that sets the need for a laser safety program. More on this in other chapters.

2.3 Terms to know

(Some of these terms are re-examined in chapter 4 dealing with beam properties, and in some cases the terms are defined by the Center for Devices and Radiological Health, US FDA.)

Accessible emission level is the magnitude of accessible laser or collateral radiation of a specific wavelength and emission duration at a particular point as measured according to CDRH procedures.

Accessible emission limit is the maximum accessible emission level permitted within a particular class per CDRH rules, same definition per IEC '*accessible emission limit (AEL)* the maximum accessible emission level permitted within a particular class'.

A *laser* is a device that produces radiant energy predominantly by stimulated emission. Laser radiation may have a high degree of spatial and temporal coherence. It is an acronym for **L**ight **A**mplification by **S**timulated **E**mission of **R**adiation.

A *laser system* is an assembly of electrical, mechanical, and optical components which includes a laser. In CDRH terms a *laser system* means a laser in combination with an appropriate laser energy source with or without additional incorporated components.

An *embedded laser* is an enclosed laser that has a higher classification than the laser system in which it is incorporated, where the system's lower classification is appropriate due to the engineering features limiting accessible emission.

2.4 Explanation of individual laser classes

There are some slight differences between laser classes and groups that define laser hazard classification. The important ones are CDRH and IEC. ANSI is a user standard and no laser manufacturer classifies its equipment per ANSI. Even if a Laser Safety Officer (LSO) reclassifies or does the initial classification of, say, a homebuilt laser, they will look at CDRH and IEC for class specifications.

I will say that unless classifying lasers is part of one's job the typical LSO should stay away from this activity. Once they classify or reclassify a laser they have taken on a certain level of responsibility for the laser or laser system.

2.5 System classification

In my mind this is different to product classification. I may have a laser set-up, which either due to limited open beam or enclosures I might define as a Class 1 system during normal operation, even if the laser source is a Class 4 laser. One can frequently come across this situation in the manufacturing environment, where a Class 4 laser is interfaced with an assembly line, where there is no exposure potential.

The following tables compare CDRH and IEC classification (table 2.1), classification output levels (table 2.2) and finally (table 2.3) shows what general safety program requirements go with each laser class.

Table 2.1. CDRH IEC comparison.

CDRH Class	IEC 60825-1	Specifications
1	1	Not an eye or skin hazard under all conditions.
NA	1M	Will not cause permanent eye damage due to aversion reflex, applies to invisible wavelengths. Not to be viewed by magnifying optics, beams are expanded to reduce irradiance to safe viewing level.
2a (IIa)	NA	Only applies to visible wavelengths, protection due to adversion reflex, not safe for continuous viewing.
2 (II)	2	Will not cause permanent eye damage, counts on aversion reflex to keep exposure to less than 0.25 s. Limited to visible wavelengths
NA	2M	Will not cause permanent eye damage due to aversion reflex, applies to visible wavelengths. Not to be viewed by magnifying optics, beams are expanded to reduce irradiance to safe viewing level.
3A (IIIa)	3R	Direct intrabeam viewing is a potential eye hazard, should be transient effect.
3B (IIIb)	3B	Direct and specular exposure has strong potential to be hazardous to the eye, generally not sufficient energy to damage skin.
4 (IV)	4	Direct and specular exposure a definite eye and skin hazard. Can be a diffuse reflection hazard if close enough to reflecting surface. Has potential to cause fire and combust gases. At high enough outputs has potential to ionize air.

Table 2.2. Laser classification output limits.

Class 1	No eye or skin hazard, a few hundred microwatts or less
Class 1M	Expanded beam, invisible, irradiance below maximum permissible exposure (MPE)
Class 2	1 mW, aversion reflex
Class 2M	Expanded beam, visible, irradiance below MPE
Class 3R	1–5 mW, safe for momentary exposure, unless viewed through collecting optics
Class 3B	Output range 5–500 mW—continuous wave, cannot produce 125 mJ in less than 0.25 s
Class 4	>500 mW—CW, 125 J/pulse in less than 0.25 s

NOTE ON: limitations

Laser safety classification relates to accessible laser radiation—this classification doesn't take into account additional hazards, such as electricity, collateral radiation, fume, noise, etc. Laser safety classification relates to normal use of the product—the classification can change during maintenance or service, or when the original device forms a part of a complex installation. Laser safety classification relates to a single product—it doesn't account for accumulative exposure from multiple sources. It is based on potential exposure during 'normal operation'.

Table 2.3. Program elements by classification.

Class	LSO	Training	Administrative control measures
Class 1	Not required	Not required	Not required
Class 1M	All application dependent, generally nothing required		
Class 2	Not required	Not required	Not required
Class 2M	All application dependent, generally nothing required		
Class 3R	Not required	Not required	Not required
Class 3B	Must have	Yes	Yes
Class 4	Must have	Oh! Yes	A must have

2.5.1 Class 1—CDRH

A laser, or laser system, which does not present a hazard to skin or eyes for any wavelength or exposure time. Exposure varies with wavelength. For ultraviolet, 0.2–0.4 µm exposure is less than from 0.8 nW to 0.8 µW. Visible light exposure varies from 0.4 µW to 200 µW, and for near IR, the exposure is <200 µW.

Class 1—ANSI

A Class 1 laser is considered 'safe' under reasonably foreseeable conditions of operation and presents no hazard to the eye or skin. A Class 1 laser may exceed accessible emission limits for Class 1 but, because of the geometrical spread of the emitted radiation, the laser does not cause harmful levels of exposure to the *unaided* eye. Laser(s) or laser systems are exempt from any laser control measures.

IEC

Class 1 laser product

Any laser product which does not permit human access to laser radiation in excess of the accessible emission limits of Class 1 for applicable wavelengths and emission durations.

CDRH

2.5.2 Class 1M

This classification is not found in the CDRH code. It was established after the laser product safety rules were issued.

Class 1M ANSI

Class 1M lasers, or laser systems, are exempt from any control measures or other forms of surveillance except when optically aided direct viewing of the beam is expected and/or during unattended operation where the beam is directed into a location where it can be directly viewed by the general public and/or personnel that may be uninformed about the hazards.

IEC

Class 1M laser product

Any laser product in the wavelength range from 302.5 nm to 4000 nm which does not permit human access to laser radiation in excess of the accessible emission limits of Class 1 for applicable wavelengths and emission durations, where the level of radiation is measured but is evaluated with smaller measurement apertures or at a greater distance from the apparent source than those used for Class 1 laser products.

Note: The output of a Class 1M product is therefore potentially hazardous when viewed using an optical instrument.

2.5.3 IEC Class 1C

Note: This classification is not, at the time of printing this book, a recognized ANSI or CDRH classification. The expectation is that it will be classified in future editions of both.

A laser product which is designed explicitly for contact application with the skin or non-ocular tissue and that, during operation, ocular hazard is prevented by engineering means. During operation and when in contact with skin or non-ocular tissue, irradiance or radiant exposure levels may exceed the skin MPE as necessary for the intended treatment procedure.

2.5.4 Class 2—CDRH

Any visible laser with an output less than 1 mW of power. Warning label requirements: yellow caution label stating maximum output of 1 mW. Generally used as classroom lab lasers, supermarket scanners and laser pointers.

Class 2 ANSI

Class 2 lasers or laser systems are exempt from any control measures except when intentional direct viewing of the beam is possible.

Class 2 IEC
2.5.5 Class 2M

This classification is not found in the CDRH code. It was established after the laser product safety rules were issued.

IEC Class 2M laser product

Any laser product in the wavelength range 400 nm to 700 nm which does not permit human access to laser radiation in excess of the accessible emission limits of Class 2 for applicable wavelengths and emission durations, where the level of radiation is measured but is evaluated with smaller measurement apertures or at a greater distance from the apparent source than those used for Class 2 laser products.

Note: The output of a Class 2M product is therefore potentially hazardous when viewed using an optical instrument.

2.5.6 Class 3R—CDRH

Any visible laser with an output over 1 mW of power with a maximum output of 5 mW of power. Warning label requirements: red danger label stating maximum output of 5 mW. Also used as classroom lab lasers, in holography, laser pointers, leveling instruments, measuring devices and alignment equipment.

Note: Classification definitions for Class 3R, Class 3B and Class 4 are much clearer if one looks at the CDRH and ANSI explanation than IEC.

Class 3R—IEC
2.5.7 Class 3R

Any laser product which permits human access to laser radiation in excess of the accessible emission limits of Class 1 and Class 2 as applicable, but which does not permit human access to laser radiation in excess of the accessible emission limits of Classes 3R and 3B (respectively) for any emission duration and wavelength.

Class 3B
IEC

Class 3B: A Class 3B laser is hazardous if the eye is exposed directly, but diffuse reflections such as from paper or other matte surfaces are not harmful. Continuous lasers in the wavelength range from 315 nm to far infrared are limited to 0.5 W. For pulsed lasers between 400 nm and 700 nm, the limit is 30 mJ. Other limits apply to other wavelengths and to ultrashort pulsed lasers. Protective eyewear is typically required where direct viewing of a Class 3B laser beam may occur. Class 3B lasers must be equipped with a key switch and a safety interlock.

2.5.8 Class 4 CDRH

Any laser with an output over 5 mW of power with a maximum output of 500 mW of power and all invisible lasers with an output up to 400 mW. Warning label requirements: red danger label stating maximum output. These lasers also require a key switch for operation and a 3.5 s delay when the laser is turned on. Used in many of the same applications as Class 3 when more power is required.

IEC Class 4: Class 4 lasers include all lasers with beam power greater than Class 3B. In addition to posing significant eye hazards, with potentially devastating and permanent eye damage as a result of direct beam viewing, diffuse reflections are also harmful to the eyes within the distance called the nominal hazard zone. Class 4 lasers are also able to cut or burn skin. In addition, these lasers may ignite combustible materials, and thus represent a fire risk, in some cases. Class 4 lasers must be equipped with a key switch and a safety interlock.

2.6 Classification changes that are being discussed

The present laser hazard classification system dates back to at least 1973 and it has served the laser community well. Classes 1M, 2M and 1C have been the newest classifications to be incorporated into the established laser classification scheme. What might be coming? Two ideas have been raised but not yet accepted.

2.6.1 Class 5

The days when people thought there would be a power output limit for lasers has long passed. Today, megawatt, terawatt and petawatt systems can be purchased from commercial firms and are well established in research facilities worldwide. Class 4 is set for a continuous wave laser as any output greater than 0.5 W. Should a megawatt laser be classified as the same? The question the standard committees ask is: does a megawatt or higher output laser have any unique controls that would not apply say for a 1000 W or for that matter 10 W laser? Training would be required for sure, engineering and administrative controls to some level would apply. The major difference is non-beam related hazards associated with these systems. Most commonly ionizing radiation based on target. Some other non-beam hazards include electromagnetic pulse (EMP), exotic coolants (chemical hazard), as well as personnel exclusion.

Part of the argument for Class 5 is the desire to alert users, or those in the area, of the extreme power output of these systems. If a Class 5 or 4A and 4B or 4L (L for lethal) will come about is hard to say, but for sure it will not be before 2023.

2.6.2 Classification based on control measures—laser control groups

The idea that classification should be based not on output but rather on what control measures are needed is also being discussed. This was first presented by Robert Aldrich of the US Navy (at least as far as this author knows). This approach would fit in well for the Class 5 advocates but would also cause laser devices to have two classifications in my mind, one for the laser itself and the other for how it is used. So one might have say a Class 4 laser but in use call it a laser control group, A, B, or C (final name and delineation to be determined) 3. Of course, levels of control groups would need to be a consensus control.

2.7 Training slides on classification

When one develops a laser training presentation, the more it relates to the audience the more effective it will be. Therefore, as you look over the following slide set/power point set, try to populate each with images that will relate to your audience. Take images at your site and from the company web site.

Following are some slides you can adopt to your training presentation:

Slide 1 Laser hazard classification

- A common question laser users hear is 'What class laser do you have?'
- The purpose of laser hazard classification is to understand the potential hazard of the laser system just by the classification without detailed knowledge of the laser
- It is, in many ways, similar to the international system for chemical hazard identification
- It is a means of hazard communication
- Most laser users never quite know the hazard classification of their lasers or what it means

Slide 2 Class 1

Power output is too low to cause eye or skin injury

Examples:

- power = few microwatts
- high-powered expanded beam

Slide 3 Class 1 Product

- A product that completely contains a laser
- During normal operation there is no potential for exposure to the laser beam
- Therefore, a kilowatt laser can be turned into a Class 1 product
- No laser safety training or laser protective eyewear is needed by operators or those around the Class 1 product
- Biotech equipment, during normal operation, are frequently Class 1 products

Slide 4 Class 1 Product concern

- As stated, during operation of a Class 1 product there is no laser hazard, no operator training, no eyewear and no signage is required
- All that changes when these systems are opened up for service
- If a potential for exposure exists then it must be dealt with
- In house staff performing service
- Outside vendor

Slide 5 Class 1M Invisible wavelengths

- Incapable of producing damaging radiation levels during operation
- Unless the beam is viewed with an optical instrument
  - Eye-loupe or telescope
- Exempt from any control measure other than to prevent potentially hazardous optically aided viewing
- Example
  - Free space communication

Slide 6 Class 2 Visible wavelengths

- Class 2 (low power)
  - Visible (400 nm–700 nm)
  - CW upper limit is 1 mW
- Counts on aversion reflex of 0.25 s to prevent eye damage
- Can be a hazard if the aversion response is over-ridden or slowed
- Examples
  - Supermarket or barcode scanners
  - Some laser pointers

Slide 7 Class 2M Visible wavelengths

- Class 2M
  - Visible (400 nm–700 nm)
- Eye protection is aversion response for unaided viewing
- Potentially hazardous when viewed with optical aid
- CW upper limit is 1 mW
- Examples
  - Leveling instruments and some construction industry lasers, interferometers

Slide 8 What does M mean?

- **M** is for magnification
- **Class 1M** laser is Class 1 unless magnifying optics are used
- **Class 2M** laser is Class 2 unless magnifying optics are used
- **M** classes usually apply to expanded or diverging beams

Slide 9 Class 3R

- Safe for momentary viewing
- As a commercial product can only be visible
- Between 1–5 mW
- Can be a hazard if viewed through optics
- Majority of legal laser pointers are 3R
- Some alignment lasers
- **R** stands for Reduced Requirements

Slide 10 Class 3B Getting serious now

- Can be visible or invisible
- Intrabeam viewing hazard
- Specular reflection hazard
- CW output between 5 mW–500 mW
- Pulse limit cannot produce 125 mJ in less than 0.25 s
- Starting point for required laser safety program

Slide 11 Class 4

- Output higher than Class 3B
- Poses greatest danger
- Fire hazard
- Skin hazard
- Intrabeam viewing hazard
- Specular reflection hazard
- Possible diffuse reflection hazard
- Can produce more than 125 mJ in less than 0.25 s
- Over 500 mW CW

Slide 12 Why is classification so important to you?

- Your entire laser safety program even if you have only one laser system is all based on what class lasers your facility has
- Notice *not* the number of lasers, but *what class laser you have*
- Let me repeat: The standards do not take a graded approach on whether you need a program or not
  - There is a graded approach on controls
  - If you have Class 3B or Class 4 laser system, or even embedded Class 3B or Class 4 lasers, you need a laser safety program and LSO

IOP Publishing

Laser Safety
Practical knowledge and solutions
Ken Barat

Chapter 3

Biological effects: something you should know about

Ken Barat

3.1 Confession time

Numerous scholarly and peer reviewed articles, chapters and texts have been written on the biological effects of laser radiation. This chapter is not trying to reproduce that or give the detail the topic deserves. The goal is to present an overview and to touch on some topics the reader may not have considered when thinking about laser radiation biological effects.

3.2 Let's be truthful

If it were not for the harmful effects of laser radiation, laser safety concerns would not exist. Let's be clear: not all laser radiation interaction with human tissue is harmful. Many uses exist for lasers in medical settings. But from a laser safety perspective it is keeping one safe from laser radiation levels that can cause unintentional injury that we are concerned with.

Laser radiation is not a 'one size fits all'. Different wavelengths have different effects on cellular components. The eye is the chief organ of concern and the wavelength range of 400 nm–1400 nm is the region of greatest potential harm. This wavelength range is known as the 'Retinal Hazard Region' the band of wavelengths, visible and near-infrared that present the greatest risk to one's vision.

3.3 Oh! I forgot about that

The more one works with a laser, the more likely one is to forget its harmful effect. It is just human nature. The more one drives the more likely you are to drive through risky weather conditions. The more one works with lasers the more likely you are to either be over confident or forget about the possible negative effects and risks.

3.4 What is all the concern over?

The critical reason why 400 nm–1400 nm (or, more precisely 380 nm–1400 nm) is such an eye hazard concern really comes down to spot size on the retina. When we look at an object the average spot size of the focus beam on the macula/fovea is 200–300 microns. However, when the lens focuses a laser beam from the retinal hazard region or wavelengths the spot size is 20–30 microns. The spot size is reduced by a factor of 10. This yields an optical gain of 100 000. Which increases the irradiance within the spot to incredible powers.

1 mW cm^{-2} at the lens is 100 000 W cm^{-2} at the macula.

Irradiance is a critical factor to understand. It is the power distribution or deposited per square centimeter. A safe exposure is the maximum permissible exposure (MPE) or easier to consider the speed limit for laser exposure. Exposure up to and including the MPE will not cause damage, just like driving at the speed limit should not get you a speeding ticket. As your speed increases over the speed limit, so does your chance of a ticket or accident. It is the same for the more exposure over the MPE one receives the greater likelihood of biological damage.

Explained that way the concept is rather simple. Complexity is added as we try to explain the damage mechanisms of different wavelength ranges. Every year additional research is carried out in order to achieve a greater understanding of these mechanisms. With that research our understanding and comprehension changes. As an example, some wavelengths that were thought to only be a retinal hazard, now seem to deposit some energy in the cornea and they are transmitted through to the retina. This new knowledge and understanding is translated into new MPE values and correction factors. That is evident by changes in MPE values over the years and new editions of laser standards.

Let's now talk about damage mechanisms in general terms, knowing that whatever is said will be modified in the next few years, but the sections below will serve as a good foundation.

3.5 All aboard—train station analogy

The lasers friend, Albert Einstein, presented a very famous *thought experiment*, that involving an observer at a train platform and a train passing by the station. Let's use that theme to explain laser wavelengths and what parts of the eye they affect. If one images a train track from the front of the eye to the back, different wavelengths get off at different stations.

You may now be wondering: what is he talking about?

Imagine you're standing on a train while your friend is standing outside the train, watching it pass by. If lightning struck both ends of the train, your friend would see both bolts of lightning strike at the same time. But on the train, you are closer to the bolt of lightning that the train is moving toward. So, you see this lightning first because the light has a shorter distance to travel. This thought experiment showed that time moves differently for someone moving than for someone standing still. This is a cornerstone in Einstein's special theory of relativity.

Some ultraviolet, mid- to far-infrared, 'get off' at the lens and some at the cornea, while visible and near-infrared, like the express train, get off at the end of the line. If the exposure takes a side spur line and misses the macula/fovea, but still hits another part of the retina the damage may easily go unnoticed and have no effect on vision.

3.6 Injury below damage threshold

People can be exposed to light levels *below* what is needed for an injury, but still *high* enough to cause temporary vision problems. The most common example of this is aircraft pilot illumination but it can happen in other circumstances.

Distraction: where the light does not obscure vision but can distract the pilot. Light level 0.5 μW cm^{-2}; for example, a legal 5 mW laser pointer at 3700 feet (1130 m), just think of the distance if one has a 5 W laser system.

Glare: where it is hard to see through the light to the background scene. Light level 5.0 μW cm^{-2}; for example, a legal 5 mW laser pointer at 1200 feet (365 m), just think of the distance if one has a 5 W laser system.

Flash blindness: where the image takes from a few seconds to a few minutes to fade away, depending on how much light entered the eye. Light level 50 μW cm^{-2}; for example, a legal 5 mW laser pointer at 350 feet (107 m), just think of the distance if one has a 5 W laser system.

3.7 Indoor problems below the MPE do exist

While one thinks of visible light distraction chiefly as an outdoor issue, it also applies to indoor applications (figure 3.1), in particular, to green light common from many pump lasers. People also complain that some green pointers are too bright for them to look at. Diffuse light from a green pump laser can light up and reflect off specular surfaces in a laser lab. Many set ups use a clear tube to keep fingers and shine objects out of this pump beam, which is much better than leaving the beam to travel in open space. It does not matter that the usual distance is only a few centimeters. What the user should do is use an opaque tube, see examples below as successful approaches (figure 3.2). While the diffuse light is rarely at a level to cause injury, it can cause eye strain, headaches and even more, the perception of risk to those who see the light. Of course a dilemma is that if one is wearing glasses for 527 nm or 532 nm you will not see the diffuse light.

Figure 3.1. Diffuse green light.

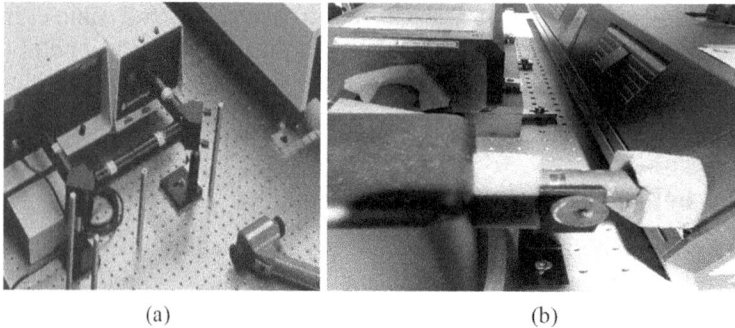

(a)   (b)

Figure 3.2. (a) Opaque beam tubes, (b) foam at tube ends to block visible scatter.

3.8 Equipment damage

To some their equipment is just as valuable as their sight. Laser beams can damage equipment. Cameras, video cameras, recording equipment with CMOS or CCD elements are susceptible to damage. Just like the eye, a direct hit onto a camera lens has the potential to damage equipment, even below the MPE level for biological concerns.

Where does that light go?

3.9 So where do wavelengths go?

3.9.1 UV

180 nm–400 nm, let me say now if you are looking for exposure value numbers below 180 nm you will not find them in any laser standard. So, if your laser is generating harmonics below 180 nm, do not waste your time looking for MPE values. UV tends to be absorbed by one's cornea and lens. The biggest problem is that there is no immediate sensation of injury, this does not occur until hours later. Just like when you are driving with your arm out of your window, it is hours later when the pain of sunburn hits, not while you were being overexposed.

In addition, just when you think you are safe, many chemicals one can come in contact with will increase your sensitivity to UV, lowering the threshold for damage (table 3.1).

3.9.2 Visible

In the visible spectrum (400 nm–700 nm) (also applies to the near-infrared region 700 nm–1400 nm), laser radiation starts its journey through the eye at the cornea (a transparent membrane), then through the aqueous humor, the pupil an opening in the iris, the lens, the vitreous humor, and several layers of blood vessels and nerve layers which form the retina. Finally, it reaches rods and cones which make up the retina.

The wavelengths of the visible spectrum determine the visible sensation of color: violet at 400 nm, red at 700 nm, and the other colors of the visible spectrum in between. The eye's perception of color is not uniform; some colors look more intense, even if all are at the same output.

Table 3.1. Photosensitizing agents.

Tetracyclines
Sulfonamides
Coal tar—dandruff shampoos
Amiodarone
Diltiazem
Quinine
Quinidine
Hydroxychloroquine
Enalapril
Dapsone
Voriconazole
Fluoroquinolones e.g. ciprofloxacin
Non-steroidal anti-inflammatory drugs (NSAIDs)
Ibuprofen
Naproxen
Ketoprofen
Celecoxib
Diuretics
Frusemide
Bumetanide
Hydrochlorothiazide

3.9.3 Near-infrared 700 nm–1400 nm

This band of wavelengths is not your friend. You cannot see it, well barely up to about 850 nm. You have no aversion response, so it will be deposited to your retina and can cause severe damage.

3.9.4 Mid- and far-IR

Part of the IR-B (mid-IR) band of wavelengths is also special. Wavelengths from 1400 nm–2600 nm can be absorbed by all the structures of the eye except the retina (mostly absorbed in the cornea). This is known as volume absorption in the eye. Denaturation of the cornea can occur. Heat can build up in the aqueous and vitreous humor and long-term effects may include cataracts. Although these wavelengths are normally invisible to humans, some victims have reported seeing a 'white flash' when overexposed to lasers in this band.

3.10 How is damage caused?

When radiation is absorbed, the effect on the absorbing biological tissue is photo-chemical, thermal, or mechanical. Ultraviolet radiation effects are primarily photo-chemical, infrared radiation effects are primarily thermal, and in the visible part of the spectrum both effects are present. A small amount of radiation is harmless, but when the intensity of the radiation is sufficiently high, the absorbing tissue will be damaged. The particularly high levels of energy impinging on tissue are the source of the concern regarding laser exposure.

Several mechanisms explain how a laser lesion comes about. The physical trauma of laser radiation may be thermal, thermo-acoustic, or photochemical. After the initial hit, biological reactions within the tissue may occur. Protein mutations that occur are related to the radiant energy and duration of the exposure. The ripping or tearing of tissue, is caused by a thermo-acoustic transient pressure wave, this is a mechanical effect.

3.11 The anatomy of your eye

3.11.1 The cornea

The cornea, the transparent bulge on the front of the eye, is the primary refracting structure of the eye. Because of the difference in refractive indexes of air and the cornea, 80 percent of the refraction of light takes place as the light enters the eye. This explains why corrective eye surgery is done on the cornea rather than attempting to reshape or replace the lens.

3.11.2 The aqueous chambers

The aqueous humor and vitreous humor allow for this movement and maintain the index of refraction through the eye. The retina is the light absorbing structure of the eye.

3.11.3 The lens

The lens is moveable/re-shapeable by the ciliary muscles and serves as the dynamic focus for the eye.

3.11.4 The retina

The visual receptors (rods and cones) are located in the retina, which contains a blind spot at the point of entry of the optic nerve into the eye. This natural blind spot is caused by the optic nerve and disk and is about 2.5 mm in diameter. Damage to the optic disk, would most likely result in a total loss of vision in that eye.

The portion of the retina that is most sensitive to detail and color discrimination is the fovea where there is a high concentration of cone cells that are responsible for higher visual acuity and color vision. The fovea is a very narrow area of the central portion of the retina and the cone of vision in the fovea is only about a two-degree area of sight.

The rod cells are most sensitive to low light levels and are distributed over the entire retina except within the foveal area. The retinal pigment epithelium is the base of the retina—critical in retinal metabolism and photochemistry—this is where retinal detachment occurs.

3.11.5 The iris/pupil

The iris serves as the variable aperture for different light levels and can change in size, in most people when they tell a lie or are attracted to someone. Pupil size by action of the iris will change based on light conditions:

- Typical sizes
  - ○ 2 mm daylight
  - ○ 3 mm indoor
  - ○ 7 mm dark adapted
  - ○ 8 mm dilated (for eye exam)

All laser safety calculations and software defaults to 7 mm pupil size, so any beam 7 mm or less is considered to deposit all its energy in your eye.

3.12 800 nm trap

Though, as the curves show, there is a very low level of sensitivity at 400 nm and 700 nm, the eye is still responsive at those wavelengths, and even beyond if the energy is intense enough. A large number of laser exposures have occurred to people working at 800 nm. Why? Well they see a faint dot (reddish), and the instinct that faint is equivalent to weak fools them into making errors of judgment. Well, at 780 nm–850 nm your eye perceives less than 1% of the photons that are present, but the 99% are still present. So one is not going to get a weak exposure, but a tidal wave of photons.

Does this image (figure 3.3) help you understand 800 nm trap? Based on the surface view how big would you say the iceberg is?

Figure 3.3. Iceberg, its true size and shape is hidden under water.

3.12.1 Damage mechanisms by wavelength

Ultraviolet radiation exposure to the eye results in photochemical reactions such as photokeratitis, erythema and photochemical cataracts. UV skin exposure can cause erythema, skin cancer, accelerated aging, increased pigmentation, photosensitive reaction and skin burn.

Visible radiation exposure to the eye results in photochemical reactions and thermal retinal injury.

Infrared radiation exposure to the eye results in cataracts, retinal burns, corneal burns, and aqueous flare. IR skin exposure can burn skin.

3.13 Things to know

3.13.1 Aversion response

Our reflex to extremely bright light is to blink and turn the head away from the source (the pupil will also constrict). This reflex is called the aversion response and takes approximately 0.25 s. The reflex does not occur with invisible wavelengths and it can only limit the amount of damage from exposures to lower power/energy lasers in the visible band.

3.13.2 Near IR effects

There is no aversion response to invisible beams. Since NIR wavelengths are focused by the eye to your fovea, they are considered the most dangerous to laser users.

Multiple strikes from a pulse laser: it is safe to say pulsed lasers have much higher energy densities per pulse than a CW beam. This is because all the energy is released in very short time intervals. The shorter pulse width beams have more power so, therefore, more damage can be done because tissues must absorb and dissipate the heat energy in shorter timeframes before the damage threshold is reached.

3.13.3 Injuries outside the fovea

A laser lesion in the peripheral areas of the retina may not be noticed or cause a significant reduction in visual functioning (i.e. visual acuity) because the vision in those areas is very poor compared to vision in the fovea. Lesions in areas other than the fovea are referred to as 'off axis' damage or hits. Most are benign. In fact, a laser may be used to surgically treat diabetes-related problems in the retina. Some diabetics suffer visual dysfunction because of weakened/leaky blood vessels in the retina. A laser can be used to cauterize the smaller vessels and visual function is improved by that procedure.

If the off axis damage is on the major nerves or blood vessels in the eye, the visual functioning of the cells downstream can result in disrupted vision in much larger areas than the initial scotoma. Blood leaking into the inner chamber of the eye from an off axis hit can also severely impact foveal vision.

3.13.4 Vitreal hemorrhages

More severe tissue reaction to the rapid buildup of heat (without being able to dissipate that heat energy) can cause mechanical destruction in the retina. The fluid in a blister on the retina can cause bleeding within the blister. The physical separation of the retinal layers by fluid and/or blood can destroy the connections between the visual cells (rods and cones) and their nerves. Another effect of blood in a retinal blister is caused by the toxic reaction of the visual cells when directly exposed to hemoglobin in the blood. The hemoglobin will kill those cells.

Mechanical destruction may occur in the form of ripping or tearing of tissue. The tearing is the result from a thermo-acoustic transient pressure valve. Blood may be released into the vitreous humor and diffuse through that medium which results in the victim's perception of having the field of view turn red if the hemorrhage obscures the fovea. This bleeding is called a vitreal hemorrhage.

Retinal tears near the fovea are particularly severe because central vision is affected and because the damage is usually permanent. The blood in the vitreous humor will eventually be absorbed or it can be surgically removed. Even though some tissue healing may occur with time, the scarring left on the retina will eventually cause tension on the delicate retinal tissues and produce a 'rippling' effect. The scarring and ripples result in reduced visual acuity. Medical experts have routinely tracked the healing process of laser accident victims who have suffered retinal tears and vitreal hemorrhages. Visual acuity of the victims is at its worst immediately following the tear. As the healing begins, some acuity is regained but the scarring then causes acuity to degrade again. Even though some visual function is restored, it never returns to the level before the damage was done.

3.13.5 Blood in the eye, toxic effects

A secondary effect is the toxic reaction of visual cells to hemoglobin. Another effect of off axis lesions on the fovea is that concerned with damage to nerve and blood tracts between the fovea and optic disk.

3.13.6 Beam size does have an effect

If the beam is smaller than the pupil, spherical aberrations and forward scattering spread the 'point' image so that the actual distribution of light from a point is larger than the theoretical image size. In general, the larger the pupil, the smaller the point spread and the greater the magnification factor or amplification of light at the retina, as compared with the intensity at the cornea. Pupil size is also an important variable when considering the ambient lighting in a laser operating area. At night, the pupil is dilated and allows as much as 100 times more light into the eye. Therefore, dark rooms or night exposures to lasers are more hazardous.

When viewing an extended source, such as the reflection of a laser from a diffuse highly reflective surface, the geometry of the situation results in an image, which is of constant brightness until a critical image size is reached, and then the brightness decreases. For long exposures, the large and small image size damage thresholds are different because of thermal conduction.

3.13.7 Damn! corneal injury hurts

Unlike the retina, the cornea has many pain nerves. Corneal damage is usually very painful. Fortunately, minor corneal injuries may heal without any long lasting effects within 24 h–48 h. More serious corneal lesions may be permanent or require transplant surgery. Lens injuries may also be treated surgically.

3.14 Physiological damage mechanisms

Several mechanisms can be involved in producing a laser lesion. The initial physical trauma of laser radiation may be thermal, thermo-acoustic, or photochemical. The initial trauma is followed by biological reactions within the tissue itself. Protein mutations that occur are related to the radiant energy and duration of the exposure. Mechanical destruction, that is the ripping or tearing of tissue, is caused by a thermo-acoustic transient pressure wave. Since the eye is made to collect and concentrate light on the retina, the eye is the organ most sensitive to visible and near IR laser radiation (400 nm–1400 nm). If the eye will be safe at these wavelengths, the rest of the body will, too.

3.15 Quick summary

180 nm–400 nm (ultraviolet) is absorbed by the cornea and lens, and the damage mechanism is 'photokeratitis'.

400 nm–700 nm (visible spectrum) visible light passes through the cornea and is focused onto the retina at the back of the eye. This macula/fovea tissue is a small zone (2%–4% of your retinal surface) made up of around 1 million cone cells and is where all critical vision takes place. The rest of the retina is primarily for motion detection. Laser damage outside the macula/fovea may have little to no adverse effect on vision.

700 nm–1400 nm (near-infrared) IR energy, like visible light, is focused on the retina, but is not interpreted into vision. However, IR energy is still absorbed by the retina and can cause injury. One of the documented effects is the causation of a bubble that then bursts, damaging your retinal cells with a 'popping' sound from inside your eye, OUCH!

1400 nm–1 mm Like UV energy these invisible wavelengths are absorbed by the cornea and lens. Since these wavelengths do not reach the retina, they are sometimes falsely termed 'eye safe'. (The term is often associated with the 1530 nm lasers used in telecommunication applications.) But energy at these wavelengths can cause eye damage!

3.16 Skin

Skin exposure. Just like the eye, different wavelengths of laser energy are absorbed by different layers of the skin. The MPE for skin is generally determined for a skin exposure of 3.5 cm^2. Counter-intuitively, the larger the area of skin exposed, the lower the MPE, since there is less surface area to draw away the heat.

Certain medications can cause an increased photosensitivity to the eye and the skin (increased reaction, sensitivity and effects from light exposure). These are collectively known as photosensitizers.

Note 1: Eyelids are the body's thinness membrane and do not provide laser eye protection.

Note 2: NIR (700–1400 nm) have the deepest penetration into the skin.

3.17 Conclusion

Do not look at laser with remaining eye.

Chapter 4

Laser safety terms: the language LSOs speak

Ken Barat

4.1 Introduction

Every discipline has its own language and laser safety is no exception. Some of the terms come from standards, other terms apply to laser safety tools and still some others trace back to determination of biological effects as well as the physical properties of lasers. This chapter will list and define terms that come from all these sources, many more terms could be listed, but at a minimum you should be familiar with these.

4.2 Glossary of terms

Administrative control measures:	Procedures, instructions, rules, methods, or work practices that implement or supplement engineering controls; also may specify the use of PPE. Administrative controls rely on user actions to be effective.
Alignment eyewear:	Has reduced optical density (OD) from full-protection eyewear. It is only for specific visible-wavelength procedures.
Aperture:	An opening, window, or lens through which a laser beam can pass or exit an enclosure or system.
Aperture label:	A label that displays the location where the beam exits the laser enclosure.
Certification label:	A label that displays that a laser is built to Federal Laser Product Safety Standards.
Coherent:	One of the three properties of laser light—all wavelengths in phase in space and/or time. The three properties of laser light are: coherent, directional, and monochromatic.
Continuous wave (CW) laser:	Laser generating a continuous output for a period of time equal to or greater than 0.25 s.
Control measure:	A means to mitigate potential hazards associated with the use of lasers. Control measures can be divided into three groups: engineering, administrative/procedural, and personal protective equipment (PPE).

Diffuse reflection:
A reflection in many directions from a rough surface. Surface irregularities are larger than the wavelength.

Direct viewing:
(also known as *intrabeam* viewing) Looking directly into a laser beam, also makes one look stupid.

Directional:
One of the three properties of laser light—of one specific direction. The three properties of laser light are coherent, directional, and monochromatic.

Direct reflection:
(more commonly known as *specular reflection*) A mirror-like reflection from a smooth, highly reflective surface.

Divergence:
The increase in beam diameter that accompanies an increase in distance from the source—laser light has very little divergence.

Electromagnetic radiation:
The flow of energy consisting of orthogonally vibrating electric and magnetic fields lying transverse to the direction of propagation.

Electromagnetic spectrum:
The entire wavelength range of electromagnetic radiation. Gamma rays, x-rays, ultraviolet, visible, infrared, and radio waves occupy various portions of the electromagnetic spectrum and differ only in frequency, wavelength, and photon energy. Laser light covers three wavelength ranges—ultraviolet, visible, and infrared.

Engineering control measures:
Control measures designed or incorporated into the laser or laser system (e.g. interlocks and shutters) or its application. These controls are designed to reduce or eliminate beam and non-beam hazards associated with laser operation. Engineering controls don't rely on user actions to be effective.

Fiber optics:
Fiber optics (optical fibers or cables) are long, thin strands of glass or plastic about the diameter of a human hair. They are used, singly or in bundles (optical cables), to transmit light signals.

Frequency:
The number of waves per unit time, usually measured in Hertz (Hz) (1/s). Frequency is the inverse of the time period T: $f = 1/T$. Additionally, frequency and wavelength are inversely related.

Full-protection eyewear:
Has sufficient optical density (OD) to attenuate the laser beam to a safe level below the MPE.

Infrared radiation (IR):
Electromagnetic radiation between the long-wavelength extreme of the visible spectrum (700 nm) and the shortest microwaves (1 mm).

Intrabeam viewing:
(also known as direct viewing) Looking directly into a laser beam, just makes one look stupid.

Interlock irradiance:
Incident power per unit area—usually expressed in watts per square centimeter.

Laser:
An acronym for **L**ight **A**mplification by **S**timulated **E**mission of **R**adiation. A device that produces radiant energy, predominantly by stimulated emission. Laser radiation is coherent, directional and monochromatic.

Laser barrier:
A device used to block or attenuate direct beams, diffuse laser radiation or stray beams. Laser barriers may also be used to establish a boundary for a laser controlled area.

Laser controlled area (LCA):	A laser use area where the occupancy and activity of those within is controlled and supervised. This area may be defined by walls, barriers, or other means. Within this area, potentially hazardous beam exposure is possible.
Laser generated air contaminant (LGAC):	May be generated when Class 3B and Class 4 laser beams interact with matter. Contaminants can be dust, chemical fumes, metallic fumes, aerosols, and aerosolized biological contaminants (viable bacteria, cellular debris, or viruses).
Laser manufacturer:	A facility may be considered a laser manufacturer if it builds or modifies a laser and then transfers it to a third party.
Laser safety officer (LSO):	A person who has authority and responsibility to monitor and enforce control of laser hazards and effect the knowledgeable evaluation and control of laser hazards. Always over-worked and under-paid.
Laser user or laser worker:	A person who performs work with lasers or laser systems.
Maintenance:	Performance of those adjustments or procedures which are to be carried out by the user to ensure the intended performance of the product.
Maximum permissible exposure (MPE):	The level of laser radiation to which a person may be exposed without adverse biological changes in the eye or skin.
Monochromatic:	A single or narrow band of wavelengths. The three properties of laser light are coherent, directional, and monochromatic.
Nominal hazard zone (NHZ):	The region around a laser within which the level of direct, reflected, or scattered radiation may exceed the MPE level.
Non-beam hazards:	All hazards arising from the presence of a laser system, excluding direct human exposure to direct or scattered laser radiation.
Normal operation:	The performance of the laser or laser system over the full range of its intended functions. Normal operation does not include maintenance or service.
Optical density (OD):	For a given wavelength, an expression of the transmittance of an optical element: the higher the optical density, the lower the transmittance. It is the logarithm to the base ten of the reciprocal of the transmittance at a particular wavelength.
Optically-aided viewing:	Viewing with an optical device such as an eye loupe, hand magnifier, microscope, binoculars, or telescope; optically-aided viewing does not include viewing with corrective eyewear or with indirect image converters.
Personal protective equipment (PPE):	Personal safety protective devices used to mitigate hazards associated with laser use (e.g. laser eye protection, protective clothing, and gloves). Use of PPE requires application of an administrative control.
Protective housing:	An enclosure surrounding a laser or laser system that prevents access to laser radiation above the applicable MPE level. A protective housing may also prevent access to electrical hazards.
Pulsed laser:	A laser that delivers energy in the form of a single pulse (less than 0.25 s), or in a train of pulses.

Pulse duration:	The duration of a laser pulse, usually measured as the time interval between the half-power points on the leading and trailing edges of the pulse.
Pulse interval:	The time between laser pulses.
Radian (rad):	A unit of angular measure equal to the angle subtended at the center of a circle by an arc whose length is equal to the radius of the circle. 1 radian $\sim57.3°$; 2π radians $= 360°$ radiance. Radiant flux or power output per unit solid angle per unit area expressed in watts per centimeter squared per steradian ($W{\cdot}cm^{-2}{\cdot}sr^{-1}$). Symbol: L.
Radiant energy:	Energy emitted, transferred, or received in the form of radiation. Unit: joules (J). Symbol: Q.
Reflectance:	The ratio of total reflected radiant power to total incident power.
Reflection:	Deviation of radiation following incidence on a surface.
Refraction:	The bending of a beam of light in transmission through an interface between two dissimilar media or in a medium whose refractive index is a continuous function of position (graded index medium).
Refractive index (of a medium):	The ratio of the velocity of light in a vacuum to the phase velocity in the medium. Symbol: n. Syn: index of refraction.
Repetitive pulse laser:	A laser with multiple pulses of radiant energy occurring in a sequence.
Retina:	The sensory tissue that receives the incident image formed by the cornea and lens of the human eye.
Remote interlock connector:	A connector which permits the connection of external controls placed apart from other components of the laser product.
Repetitive pulse laser:	A laser with multiple pulses of radiant energy occurring in a sequence.
Retinal hazard region:	400 nm–1400 nm—visible and near-IR wavelengths that principally affect the retina.
Safety interlock:	An automatic device associated with the protective housing of a laser product to prevent human access to Class 3 or Class 4 laser radiation when that portion of the housing is removed.
Scanning laser radiation:	Laser radiation having a time-varying direction, origin or pattern of propagation with respect to a stationary frame of reference.
Service:	The performance of procedures, typically defined as repair, to bring the laser, laser system or laser product back to full and normal operational status.
Specular reflection:	A mirror-like reflection from a smooth, highly reflective surface, surface irregularities are smaller than wavelength.
Standard operating procedures (SOPs):	Formal written description of the safety and administrative procedures to be followed in performing a specific task.
Startle response:	The involuntary response movement to a sudden unexpected stimulus.
Startle hazard:	An event such as a bright light flash that evokes a startle response.

Ultraviolet radiation (UV):	Electromagnetic radiation with wavelengths shorter than those of visible radiation, including UV-C (180 nm–280 nm), UV-B (280 nm–315 nm), and UV-A (315 nm–400 nm).
Visible light transmission (VLT):	The percentage of visible light transmitted through a filter, weighted for the response of the human eye. This is a vital parameter when selecting laser eyewear.
Visible radiation:	Electromagnetic radiation that can be detected by the human eye, specifically wavelengths which lie in the range of 400 nm–700 nm.
Wavelength:	The distance between the peaks, or crests, of a light wave, usually measured in nanometers or micrometers. The wavelength of a light wave is equal to the velocity of the light wave divided by the frequency of the light wave.

Chapter 5

Risk assessment for lasers

Randy Paura

5.1 Purpose

This is a summary risk assessment guide pertaining to laser user safety, drawn from and expanded upon the hazard assessment provided in ANSI Z136.9-2013.

5.2 Applicability

The structure and organization of the ANSI Z136.9 consensus standard for the *Safe Use of Lasers in Manufacturing Environments* is premised upon an assessment of risks for hazard groups with respective safety control measures, identified in tables 10 and 11, providing the foundation for a well-defined risk assessment process.

5.3 Preface

The term 'risk assessment' is seeing increased use in various sectors and has been realized in various forms. This chapter addresses the topic of risk assessment for a laser-based project in manufacturing environments. Its strategies can be translated to other sectors.

When it comes to occupational health and safety with lasers, risk assessment can have many forms in terms of its documentation, ranging from qualitative to quantitative, though the principles remain the same. Inherent hazards must be known, their potential risks evaluated and prioritized for elimination or mitigation with safety control measures and reassessed to determine that any residual risk of a hazard is as low as reasonably practicable (acceptable). The core of this structured process is known as risk assessment, its implementation is known as risk management.

An organization's resources, like opportunities, are valuable: they are not to be wasted. Risk assessment allows an organization to efficiently execute its responsibility for the general duty of care to its employees and affected personnel. The documentation produced through a risk assessment process will demonstrate compliance with regulatory workplace safety norms, of which ANSI Z136.9 is a recognized consensus standard for the safe use of lasers in manufacturing environments.

This chapter provides context for the LSO involved with leading a risk assessment for a laser-based project or as part of a team on a project involving lasers. Summary guidance is provided on risk assessment terms, principles and format.

5.4 Background

For traditional laser systems, ensuring conformance to regulatory build standards and validating safe use in accordance to ANSI Z136 will achieve functional safety and fulfill the general duty of care obligation by the employer to the operators and employees.

The structure of the ANSI Z136 consensus standard series for the safe use of lasers consists of:

- Determine the hazard class of a laser.
- Apply the respective control measures, summarized in tables 10 and 11 of Z136.
- Monitor and audit the safety control measures.

Embedded within this mature classification scheme is the risk assessment for the laser's capability of injuring personnel or interfering with task performance. Hence, for many laser systems in manufacturing environments, a new risk assessment methodology need not be invented from scratch or 'first principles', rather, working through the structure provided by Z136 towards the desired hazard class assesses the risk for the use of the subject laser. It is important to note that Class 4 lasers start at 0.5 W of continuous wave power and there is no subsequent breakpoint for a higher hazard class based upon a 'next level' bio-effect threshold. Within Class 4, increasing laser power/energy levels represent an increasing scale of those hazards which require commensurate safety control measures.

Below is a simplified version of a laser hazard classification scheme, referencing both ANSI Z136 and IEC 60825 series (figure 5.1).

Hazard Class	Long-term exposure		Short-term (accidental) exposure			
Hazard Class	Eye Magnified	Eye Intra-beam	Eye Magnified	Eye Intra-beam	Eye Diffuse	Skin Exposure
Class 1	Safe	Safe	Safe	Safe	Safe	Safe
Class 1M	At Risk	Safe	At Risk	Safe	Safe	Safe
Class 2	At Risk	At Risk	Safe	Safe	Safe	Safe
Class 2M	At Risk	At Risk	At Risk	Safe	Safe	Safe
Class 3R	At Risk	At Risk	Some Risk	Some Risk	Safe	Safe
Class 3B	At Risk	At Risk	At Risk	At Risk	Some Risk	Some Risk
Class 4	At Risk	At Risk	At Risk	At Risk	At Risk	At Risk

Figure 5.1. Laser hazard classification scheme—simplified.

Understanding risk assessment will allow for greater appreciation and utilization of the hazard classification scheme of ANSI Z136 and in generating the appropriate documentation. More importantly, it will enable the suitable depth and scope of a written risk assessment for those lasers and their applications as warranted.

The material presented herein is a synthesis of risk assessment principles and practices for the competent and responsible LSO to ensure that all reasonably foreseeable risks are addressed, ensuring **L**ight is **A**pplied **S**afely, **E**fficiently and **R**eliably[1].

5.5 What is a risk assessment?

Risk assessment is a term used to describe the portion of the risk management process or method. Risk management consists of:

- Identification of hazards and their factors that have the potential to cause harm (hazard identification);
- Analysis and evaluation of the potential for these hazards to be realized (risk analysis);
- Prioritization of the risks from those needing the greatest attention for safety control measures to the least (risk evaluation);
- Determination of appropriate ways to eliminate the hazard, or control the risk when the hazard cannot be eliminated (risk control);
- This process is iterative until the residual risk is acceptable.

In many safety circles, including those agencies responsible for workplace safety, risk assessment is used interchangeably for risk management. For the purposes of this chapter, the topic of risk analysis and risk control will also be covered as these are crucial for the risk assessment component of an organization's risk management process.

Risk management is the structured examination to identify those processes, operations, equipment, situations, environment, etc that may cause harm, to personnel or property. After identification of potential hazards are made, next is the determination of how severe and likely that hazard may be realized, the product of these factors is referred to as the risk. With the evaluations complete, the hazards and their respective risks evaluated, will allow for prioritization and re-assessment with safety control measures applied until the residual risk(s) are as low as reasonably practicable and acceptable. The objective is to effectively eliminate or control the harm from happening.

5.6 Fundamental concepts

Risk and hazard are used interchangeably by some. It is true that these two terms are related, but they have very distinct definitions.

---

[1] Acronym attributed to Dr D Sliney.

A hazard is what can cause harm or injury to a person or property. There are two major groups of hazards associated with lasers or equipment containing lasers: beam and non-beam hazards.

Risk is the likelihood of an occurrence of a hazardous event or exposure and the severity of injury or ill health that can be caused by the event or exposure. For the technically inclined, at its most basic, risk is a function of two variables:

Risk = f(severity of the hazard, likelihood or probability of occurrence of harm).

Lastly, the concept of acceptable risk needs to be understood. In broad terms, acceptable risk is that which has been reduced to a level that can be tolerated by the organization having regard to its legal obligations and its own OHS policy [1]. A commonly referenced objective for risk is that it be as low as reasonably acceptable (ALARA) or as low as reasonably practicable (ALARP). One reference for determining when a risk is ALARA/ALARP is where additional safety control measures do not bring a corresponding reduction in the risk valuation for a hazard. Just because one has determined that ALARA/ALARP has been achieved, the final question remains, is it acceptable?

Whereas this chapter is intended to provide summary guidance on this topic of risk assessment for lasers, the reader is encouraged to reference the more thorough consensus standard(s) that may apply for their environment and application. These include but are not limited to:

Dealing with principles:

- ISO/IEC Guide 51:2014, *Safety aspects—Guidelines for their inclusion in standards.*
- ANSI B11.0-2010, *Safety of Machinery—General Requirements and Risk Assessment.*
- ISO 12100:2010, *Safety of machinery—General principles for design—Risk assessment and risk reduction.*

Providing application or functional requirements:

- ANSI/RIA R15.06-2012 *American National Standard for Industrial Robots and Robot Systems—Safety Requirements.*
- IEC 61508 series *Functional Safety of E/E/PE.*
- IEC 62061 *Safety of machinery: Functional safety of electrical, electronic and programmable electronic control systems.*
- ISO 13849 series *Safety of machinery—Safety-related parts of control systems.*
- MIL STD 882E *USA Department of Defense—Standard Practice—System Safety.*

5.7 Terms and definitions

Within the framework of Occupational Health and Safety, the following terms need to be clearly understood:

Hazard—the potential to cause harm—which can include substances or machines, processes, methods of work or other aspects of an organization.

Risk—the likelihood that the harm from a particular hazard is realized.

Hazard identification—the process of finding, listing, and characterizing hazards.

Risk analysis—a process for comprehending the nature of hazards and determining the level of risk.

- Risk analysis provides a basis for risk evaluation and decisions about risk control.
- Information can include current and historical data, theoretical analysis, informed opinions, and the concerns of stakeholders.
- Risk analysis includes risk estimation.

Risk evaluation—the process of comparing an estimated risk against given risk criteria to determine the significance of the risk. It enables prioritization for the implementation of safety control measures.

Risk control—actions implementing risk evaluation decisions, which includes safety control measures.

- Risk control can involve monitoring, re-evaluation, and compliance with decisions.

In dealing with the likelihood and consequence of a hazardous event (occurrence), it is important to be mindful that different sectors may have different grades, as will be explained later. For now, the most basic of grading is provided below.

5.8 Likelihood of occurrence (probability)

- Very likely: Near certain to occur.
- Likely: May occur.
- Unlikely: Not likely to occur.
- Remote: So unlikely as to be near zero.

5.9 Consequence (severity)

- Catastrophic: Death or permanently disabling injury or illness (unable to return to work). Includes loss of vision in both eyes, loss of multiple limbs, quadriplegic.
- Serious: Severe debilitating injury or illness requiring more than first aid (able to return to work at some point). Includes loss of vision in one eye, loss of a limb.
- Moderate: Significant injury or illness requiring more than first aid (able to return to same job).
- Minor: No injury or slight injury requiring no more than first aid (little or no lost work time).

Using these definitions, a risk matrix can be constructed, whereby the hazard potential can be evaluated.

5.10 Why is risk assessment important?

Risk assessments are very important as they form an integral part of an occupational health and safety management plan. They help to:
- Create awareness of hazards and risk.

- Identify who may be at risk (e.g. employees, cleaners, visitors, contractors, the public, etc).
- Determine whether a control program is required for a particular hazard.
- Determine if existing control measures are adequate or if more should be done.
- Prevent injuries or illnesses, especially when done at the design or planning stage.
- Prioritize hazards and their corresponding safety control measures.
- Meet legal requirements where applicable.
- Demonstrate 'due diligence' in meeting the 'general duty' clause.

Increasingly, an end customer can require a risk assessment from the provider of a laser system, even if it is not a regulatory requirement.

5.11 Where are risk assessments noted/required by regulations?

It is a mistake to assume that since a regulation does not explicitly spell out a risk assessment for an application/process/equipment, that a risk assessment is not needed or required. Like any safety regulation, what has been codified is not normally the product of foresight, rather it is in response to an accident. Risk assessments, when identified in regulations are addressing those most prominent hazards that have had a tragic event. Demonstration of due diligence in fulfilling the obligations of the general duty clause is best achieved with a documented risk assessment.

Within the US, risk assessments are mandated in OSHA 29 CFR 1910.119, Process Safety Management of Highly Hazardous Chemicals, and EPA 40 CFR Part 68, Risk Management Plan. Of interest is where OSHA prescribes the principle of ensuring that the merit, suitability and specification of PPE against a hazard be addressed with a hazard assessment:

1910.132(d)(1)
The employer shall assess the workplace to determine if hazards are present, or are likely to be present, which necessitate the use of personal protective equipment (PPE)...

1910.132(d)(2)
The employer shall verify that the required workplace hazard assessment has been performed through a written certification that identifies the workplace evaluated; the person certifying that the evaluation has been performed; the date(s) of the hazard assessment; and, which identifies the document as a certification of hazard assessment.

Whereas OSHA requires a written *certification* that the hazard assessment has been performed, it does not prescribe that the hazard assessment itself be in writing. To fulfill the spirit, intent and letter of the law, a risk assessment would:
- Determine the inherent hazard(s) and their risk(s).
- Assess the effectiveness of safety control measures in place.

- Evaluate the residual risk(s) posed to personnel.
- Identify if PPE is warranted and enable their specifications and applicability.

Whether knowingly or not, employers make a risk assessment regarding the health and safety of their employees and affected personnel when determining the level of protective measure put in place for the operation of equipment and its processes. Due diligence is demonstrated if the employer can present some form of documented risk assessment pertaining to its equipment and processes. Remember, due diligence is what counts *before* there is an accident or incident.

It is advisable that risk assessment should be done on a periodic basis (such as a yearly safety audit) or when a change is introduced in the workplace, such as with the introduction of new (or alteration of) equipment or procedures. This is encompassed with the laser safety audits noted in the normative portions of ANSI Z136.1:2014 (A1.2)(l); Z136.8:2012 (A2.2)(m); Z136.9:2013 (A1.2)(l).

5.12 What is the goal of risk assessment?

The goal of the risk assessment process is to eliminate the unknowns affecting workplace safety, through determination of:
- What are the inherent hazards?
- What are their risks?
- How effective are the control measures?
- What are the residual risks to personnel?

This is accomplished in an organized fashion, answering the following:
(a) The object/goal is to enable knowledge within the workplace of affected personnel and stakeholders of what hazards exist, how they are contained and controlled, and what the residual risks are. An informed workplace will be able to answer the following questions:
(b) What can happen and under what circumstances?
(c) What are the possible consequences?
(d) How likely are the possible consequences to occur?
(e) Who is at risk (exposure base/population)?
(f) Is the risk controlled effectively, or is further action required?
(g) Is the risk as low as reasonably practicable (ALARP)?
(h) Is the risk acceptable, per regulations and facility OHS policy?

5.13 What is acceptable risk?

By now, this question should have come to mind, and there is considerable history behind the legal and societal evolution of acceptable risk. This subject alone has seen many attempts to identify, understand and quantify. The following is provided as baseline guidance for appreciation, as acceptable risk is relative to its environment, subject hazard of interest, and whether the risk is voluntary (such as the driver of a vehicle for transport, e.g. a car) or involuntary (such as a passenger in a vehicle, e.g. an airplane).

Zero risk is the ideal, altruistic objective for any undertaking with an inherent hazard or potential for failure. In practice, risk is inherent with any human activity or undertaking. Our appreciation for risk is based upon our experiences and perception, hence it can be problematic to define and quantify.

The World Health Organization has tackled this concept, when attempting to understand society's definition of what is an acceptable risk when it comes to their health[2].

To provide context, in a general sense, a risk is acceptable when:
- It is as low as reasonably acceptable:
  - falls below an arbitrarily defined threshold (e.g. agreed upon by affected stakeholders);
  - falls below a level already established and accepted (by custom, practice, norm);
  - falls below a comparative probability threshold that is accepted (e.g. a disease burden in a community/society).
- It is as low as reasonably practicable:
  - the cost of reducing the risk would exceed the costs saved;
  - the cost of reducing the risk would exceed the costs save when the 'costs of suffering' are also accounted for;
  - the opportunity costs would be better invested on other, higher priority health problems.
- Recognized professionals say it is safe.
- The general public is in acceptance, or there is consent by silence (non-objection).
- Governing authorities deem it acceptable.

Given that ANSI Z136.9 is an internationally recognized consensus standard for the safe use of lasers in the manufacturing environments, it is possible to draw upon and expand the format provided in tables 10 and 11 to generate a risk assessment template addressing user safety.

This supports but does not supersede the applicable CDRH or IEC 60825 compliance filing and documentation processes, which address equipment build safety. Other non-laser equipment build and user safety standards and regulations may apply.

The following worksheets are:
- One-page system laser hazard class summary (residual and inherent).
- Five-part detailed worksheets, following Z136.9 control measures groups:
  - Engineering (for laser beam related hazards);

[2] Chapter 10, Acceptable Risk from *Water Quality: Guidelines, Standards and Health* provides an appropriate framework for understanding this topic. © 2001 World Health Organization (WHO). *Water Quality: Guidelines, Standards and Health*. Edited by Lorna Fewtrell and Jamie Bartram. Published by IWA Publishing, London, UK. ISBN: 1 900222 28 0.

- ○ Administrative and procedural (for laser beam related hazards);
- ○ Personal protective equipment (for laser beam related hazards);
- ○ Miscellaneous (for laser beam related hazards);
- ○ Non-beam hazards (for secondary beam hazards and topical equipment considerations).

Color coding is also employed to provide visual guidance on the laser risk/hazard classes.

Laser Hazard Class	4	3B	3R (SLAs)	2	1
Risk/Hazard Potential	High	Moderate	Low	Low	None (ALARA)

These worksheets have been developed for industrial laser systems which utilize a high-power Class 4 laser for materials processing, embedded within a Class 1 system, to enable safe use and operation.

The text in greyscale *italic* font are for illustrative or reference purposes. The reader is referred to ANSI Z136.9 standard regarding the technical terms and assessments noted.

Manufacturer:	*System integrator. Design and build.*	*Residual Hazard Class: 1*
LMP Type:	*Class 4 laser brazing power supply, embedded into a Class 1 system/enclosure*	
Station/Project ID:	*Brass/Asset Tag, Operation Number, Station Name*	
	Ref. Laser Safety Data Sheet for details	
Laser Power Source:	*HyperDiodeLaser GmbH, Germany, HDL 6000-60*	Inherent Hazard Class: 4
Model	*HDL 6000-60*	
S/N	*WYSIWYG-314159*	
Power, Mode	*6600 Watts, Continuous Wave*	
Wavelength	*980-1060 nm, ± 10 nm, NIR, Invisible*	
MPE	*3.63 x10⁻³ W·cm⁻², t=30,000 seconds. Most restrictive. Eye*	
OD	*7: Calculated as 6.67, simplified method, most restrictive. Eye*	
NHZ_fiber-break	*N/A. Safety interlocked*	
NHD_intrabeam	*N/A. Safety interlocked focus optic assembly.*	
NHD_lens-on-laser	*228 m*	
NHD_diffuse	*7.21 m*	
Pilot Laser	*HDL, coaxial, visible for path programming*	Inherent Hazard Class: 2
Power	*1 mW, Continuous Wave, Class 2*	
Wavelength	*650 nm, Visible (red)*	
MPE	*1.00 x10⁻³ W·cm⁻², t=30,000 seconds. Most restrictive.*	
OD	*1: If viewed longer than 0.25 seconds (overcoming aversion response)*	
Process Monitoring	*HDL, laser process inspection, Class 4*	Inherent Hazard Class: 4
Power	*6 W, Continuous Wave and Pulsed (modulated)*	
Wavelength	*808 nm, Invisible*	
MPE	*1.64 x10⁻³ W·cm⁻², t=30,000 seconds. Most restrictive.*	
OD	*4: Calculated as 3.98, simplified method, most restrictive. Eye.*	
NHZ_fiber-break	*N/A. Safety interlocked*	
NHD_intrabeam	*N/A. Safety interlocked focus optic assembly.*	
NHD_lens-on-laser	*72 m*	
NHD_diffuse	*0.34 m*	

5.14 Note on structure of the risk assessment worksheets

The hazard classification scheme of the ANSI Z136 consensus standard series for the safe use of lasers, provides means to determine the scale of a laser's potential for harm between four major groups with their respective control measures to be applied. There are subdivisions within a group according to the hazard potential under certain conditions, for the purposes of industrial materials processing lasers, the subject interest is in taking the most dangerous (Class 4 lasers) with sufficient control measures so that no hazards (Class 1 conditions) are presented to personnel.

- Class 1 lasers are considered incapable of causing a hazard.
- Class 2 lasers operate in the visible spectrum, are considered a low risk hazard, with protection afforded by the human aversion response.
- Class 3R lasers are low risk lasers used in a controlled environment, such as for surveying, levelling or alignment purposes.
- Class 3B lasers are considered moderate risk hazard for intrabeam or specular reflection exposure conditions, but not necessarily diffuse exposures.
- Class 4 lasers are considered are most hazardous for direct, specular or diffuse exposure conditions, capable of being a fire ignition source. The scale and immediacy of the hazard capability of these lasers are related to their output energy or power.

As a summary guide dealing with laser beam hazards and their control measures, the general form of a risk assessment structure is provided following. It uses the scheme of dealing with an inherent Class 4 laser (associated) hazard, the application of a control measure (standard/regulation) if applicable, to achieve Class 1 condition for the operator/personnel where no further protective measures are required (other than proper training and testing for competency prior to use). Although generalized for industrial laser materials processing, the equivalency can be treated as:

Hazard/Risk Class	Hazard/Risk Level	Visual Code
4	High	
3B	Moderate	
3R (SLAs)	Low	
2	Low	
1	None, ALARP/ALARA	

The reference risk assessment worksheet is premised upon using ANSI Z136.9 for the hazard classification and safety control measures. The hazard class of the inherent laser used for materials processing (such as cutting, welding, cladding) being Class 4 (for any laser $\geqslant 0.5$ W of CW power).

For each hazardous element identified, the noted safety control measure can achieve the requisite objective of Class 1 condition: if sufficiently sized for durability and reliability. It may be that one control measure safeguards against multiple hazards. Unless warranted for redundancy requirements, it is not necessary to have multiple layers of protection against a hazard, where one will suffice.

The worksheet does provide the means to account for the nominal risk assessment factors of:

- Severity.
- Exposure.
- Avoidance.

These factors are retained to allow for larger organizations to collect a database for those higher value or unique systems that require a more rigorous assessment. But these do not need to be completed for most industrial laser applications, as Z136.9 has already detailed the hazards and their respective control measures to achieve safety. It is a matter of working through each element ascertaining if the noted control measure can be applied, or if it is addressed by another. Documenting that the residual hazard level is Class 1 for each hazardous element with the appropriate control measure will ensure that Class 1 conditions for the system as a whole are achieved.

ANSI Z136.9 identifies for each inherent hazard of a Class 4 laser, whether that safety measure is required (identified as a 'shall' in the Standard) or at least be considered (identified as a 'should' in the Standard).

The reference worksheet is drawn from table 10 in ANSI Z136.9, with the noted clause and a description of the control measure to address the inherent laser beam hazards. The LSO can document the control measure solution applied, where text in italics provides expanded notes on what the control measure is to achieve in principle or what is typically employed in industrial settings. What is presented in the residual hazard columns is for reference only to illustrate what a completed form can be.

Table 11 in Z136.9 provides for area warning signage and safety labelling requirements. These are minimal for Class 1 laser systems, as they do not pose a hazard to personnel.

Risk Assessment Worksheet													
Engineering Control Measures – Part 1/5			Inherent Hazard/Risk						Residual Hazard/Risk				
Z136.9 Clause	Description	Controls Required 4	Severity S1 / S2	Exposure E1 / E2	Avoidance A1 / A2	Risk Index (Hazard Class)	Control Measure Solution	Severity S1 / S2	Exposure E1 / E2	Avoidance A1 / A2	Risk Index (Hazard Class)	Hazard Class 1
4.4.2.1	Protective Housing	X				4	Engineered safety enclosure for containment of laser beam paths and fields.				1	✓
4.4.2.1.1	Without Protective Housing	X				4	LSO shall establish Alternative Controls as noted in written LSP				1	✓
4.4.2.1.2	Walk-in Protective Housing	X				4	Alt. control measures employed per established industrial practice, compliant to ANSI/RIA 15.06				1	✓
4.4.2.1.3	Interlocks on Removable Protective Housings	X				4	Controls reliable safety interlocks employed on human access doors and interchange dial table				1	✓
4.4.2.1.4	Service Access Panel	X				4	Service access panels are properly labeled and require a tool for removal				1	✓
4.4.2.2	Key Control	•				4	Certified laser employs requisite key control and is incorporated into the system safety interlock circuit. Additional key control not required.				N.R.	✓
4.4.2.3	Viewing Windows, Display Screens and Diffuse Display Screens	X				4	Assure viewing limited < MPE, per documented laser rated portals with required OD				1	✓
4.4.2.5	Protective Barriers and Curtains	•				4	Controlled entry into the Class 4 is such that exposure < MPE and containment barriers are designed for intended use and reasonably foreseeable single fault failure conditions				1	✓
4.4.2.6	Collecting Optics	X				4	No collecting optics are used in this system. Digital monitoring employed.				N.A.	✓
4.4.2.7.1	Fully Open Beam Path	X NHZ				4	Fully contained and controlled beam path.				N.A.	✓
4.4.2.7.2	Limited Open Beam Path	X NHZ				4	Fully contained and controlled beam path.				N.A.	✓
4.4.2.7.3	Enclosed Beam Path	X				4	Further controls not required if 4.4.2.1 and 4.4.2.1.3 are fulfilled				1	✓
4.4.2.7.4	Remote Interlock Connector	X				4	Certified laser employs requisite remote interlock connector and is incorporated into the system safety interlock circuit. Additional RIC not required.				N.R.	✓
4.4.2.7.5	Beam Stop or Attenuator	X				4	Not required for the Class 1 system, where human access to Class 4 conditions are prevented				N.R.	✓
4.4.2.8	Area Warning Device	X				4	Not required for the Class 1 system, where human access to Class 4 conditions are prevented				N.R.	✓
4.4.2.9 and 4.4.3.5	Class 4 Laser Controlled Area	X				4	Laser emission warning not required for the Class 1 system, where human access to Class 4 conditions are prevented				N.R.	✓

LEGEND:

X Shall — No requirement NHZ Nominal Hazard Zone analysis required

• Should LSP Laser Safety Program ∇ Shall if enclosed Class 3B or Class 4

LSO Laser Safety Officer Determination

N.A. Not Applicable, addressed by other mutually exclusive criteria (e.g. 4.4.2.7.1 vs 4.4.2.7.2 vs 4.4.2.7.3)

N.R. Not Required, for the residual hazard class, addressed through integration efforts as noted

✓ See supporting documentation for conformance assessment and information

Risk Assessment Worksheet												
A&P Control Measures – Part 2/5		Inherent Hazard/Risk						Residual Hazard/Risk				
		Controls Required	Severity	Exposure	Avoidance	Risk Index (Hazard Class)		Severity	Exposure	Avoidance	Risk Index (Hazard Class)	Hazard Class
Z136.9 Clause	Description	4	S1 / S2	E1 / E2	A1 / A2		A&P Measure Solution	S1 / S2	E1 / E2	A1 / A2		1
1.3.1	Written Laser Safety Program	X				4	A written laser program has been developed and implemented for the safe operation, maintenance and service of the Class 1 system with an embedded Class 4 laser.				1	✓
4.4.3.1	Standard Operating Procedures	X				4	Work instructions in place for operation, with notation on authorized users and maintenance personnel.				1	✓
4.4.3.2	Output Emission Limitations	LSO				4	Limitation on output emission of the laser determined for production requirements and quality objectives. For operator inspection and maintenance, protocols developed for the LSP and SOPs which employ minimum necessary output emission to accomplish the task at hand. Where possible, use of the				1	✓
							alignment laser, or appropriate LOTO followed to minimize or avoid human presence in the NHZ.					
4.4.3.3	Education and Training	X				4	Safety awareness training provided for Operators, safety training for maintenance personnel is commensurate with hazard Class exposure conditions.				1	✓
4.4.3.4	Authorized Personnel	X				4	Authorized personnel for operation, maintenance and service established by the LSO with posting at the system.				1	✓
4.4.3.5	Indoor Laser Controlled Area	X NHZ				4	The Class 1 system enclosure serves as the NHZ during operation. No further area control measures required.				1	✓
4.4.2.9, 4.4.3.5	Class 4 Laser Controlled Area	X				4	The Class 1 system enclosure serves as the NHZ during operation. No further area control measures required.				1	✓
4.4.3.5.2.1	Controlled Operation	•				4	Written Laser Safety Program establishes safe operation, maintenance and service events for the embedded Class 4 laser.				1	✓
4.4.3.6	Outdoor Control Measures	X NHZ				4	Intended use does not allow for outdoor operation, maintenance or service of this system.				1	✓
4.4.3.6.2	Laser in Navigable Airspace	•				4	Intended use does not allow for outdoor operation, maintenance or service of this system.				1	✓
4.4.3.7	Spectators	X				4	This is a controlled manufacturing environment. General safety awareness training required for ancillary personnel and visitors. Class 1 system does not pose a hazard to personnel.				1	✓
4.4.3.8	Alignment Procedures	X				4	Alignment procedures established for operators and maintenance.				1	✓

4.4.3.9	Service Personnel	LSO				4	Service by OEM (integrator or laser supplier) authorized personnel. Service personnel are to be laser safety trained as identified by the LSP, noted as such in purchase orders for service.				1	✓
4.4.3.10	Temporary Laser Controlled Area	—				4	Temporary LCA may be required for service, determined on an as-needed basis. LSO to review and approve.				1	✓
							Ref. written Laser Safety Program for additional information on above.					

LEGEND: **X** Shall   — No requirement   **NHZ** Nominal Hazard Zone analysis required

   • Should   **LSP** Laser Safety Program   ∇ Shall if enclosed Class 3B or Class 4

   LSO Laser Safety Officer Determination

   N.A. Not Applicable, addressed by other mutually exclusive criteria (e.g. 4.4.2.7.1 vs 4.4.2.7.2 vs 4.4.2.7.3)

   N.R. Not Required, for the residual hazard class, addressed through integration efforts as noted

   ✓ See supporting documentation for conformance assessment and information

Risk Assessment Worksheet												
PPE Control Measures – Part 3/5		Inherent Hazard/Risk						Residual Hazard/Risk				
		Controls Required	Severity	Exposure	Avoidance	Risk Index (Hazard Class)		Severity	Exposure	Avoidance	Risk Index (Hazard Class)	Hazard Class
Z136.9 Clause	**Description**	**4**	S1 / S2	E1 / E2	A1 / A2	4	**PPE Measure Solution**	S1 / S2	E1 / E2	A1 / A2	**1**	
4.4.4.2	Eye Protection (Laser Protective Eyewear)	X				4	Class 1 system does not require laser rated eyewear for operators. As determined in the LSP, maintenance and service conditions have appropriate eyewear established.				1	✓
4.4.4.3	Skin Protection	•				4	Class 1 system does not require laser skin protection for operators. Maintenance and service conditions have protocols in place which do not place personnel in the diffuse hazard zone; or where intrabeam or specular reflections are reasonably foreseeable. See written LSP.				1	✓
4.4.4.1, 4.4.4.3, 4.4.4.3.1	Protective Clothing	X				4	Class 1 system does not require laser rated protective clothing. Maintenance and service conditions have protocols in place				1	✓
							which do not place personnel in the diffuse hazard zone; or where intrabeam or specular reflections are reasonably foreseeable. See written LSP.					

LEGEND: **X** Shall   — No requirement   **NHZ** Nominal Hazard Zone analysis required

   • Should   **LSP** Laser Safety Program   ∇ Shall if enclosed Class 3B or Class 4

   LSO Laser Safety Officer Determination

   N.A. Not Applicable, addressed by other mutually exclusive criteria (e.g. 4.4.2.7.1 vs 4.4.2.7.2 vs 4.4.2.7.3)

   N.R. Not Required, for the residual hazard class, addressed through integration efforts as noted

   ✓ See supporting documentation for conformance assessment and information

Risk Assessment Worksheet													
Misc. Control Measures – Part 4/5		Inherent Hazard/Risk						Residual Hazard/Risk					
Z136.9 Clause	Description	Controls Required 4	Severity S1 / S2	Exposure E1 / E2	Avoidance A1 / A2	Risk Index (Hazard Class)	Misc. Measure Solution	Severity S1 / S2	Exposure E1 / E2	Avoidance A1 / A2	Risk Index (Hazard Class)	Hazard Class 1
4.5.1	Demonstration with General Public	X				4	This is a controlled access manufacturing environment. Class 1 system does not pose a hazard to personnel during operation. See written LSP.				1	✓
4.5.2	Laser Optical Fiber Transmission Systems	X				4	High power laser(s) for materials processing are either: - safety interlocked - have a protective housing - be risk assessed for fault exclusion per 7.3 ISO 13848-1:2015				1	✓
4.5.3	Laser Robotic Automated Installations	X NHZ				4	Robotic and automation consensus standards applied per: - ANSI/RIA 15.06 - ANSI B11.19, B11.20, B11.21, B11.25				1	✓
4.6	Laser Controlled Area Warning Signs	X				4	Appropriate area warning signs established per ANSI Z535 for the Class 1 system as identified in Z136.9, Tables 11. Posing no hazard, area warning signage requirements are minimal.				1	✓

LEGEND:

X Shall — No requirement NHZ Nominal Hazard Zone analysis required

• Should LSP Laser Safety Program ∇ Shall if enclosed Class 3B or Class 4

LSO Laser Safety Officer Determination

N.A. Not Applicable, addressed by other mutually exclusive criteria (e.g. 4.4.2.7.1 vs 4.4.2.7.2 vs 4.4.2.7.3)

N.R. Not Required, for the residual hazard class, addressed through integration efforts as noted

✓ See supporting documentation for conformance assessment and information

Consideration for intended use operational life cycle, and reasonably foreseeable fault conditions will determine appropriate design and build of the laser system for robustness, reliability and durability. Cumulative effects and/or limit states encountered during equipment operation can enable or generate potential second-order effects and hazards. The list of non-beam hazards below are not exhaustive, but serve the LSO to ensure appropriate disciplines, consensus standards and regulations are consulted.

As a summary guide dealing with Non-Beam Hazards (NBHs), the risk assessment structure is further **simplified**, but still retains the scheme of dealing with an inherent Class 4 laser (associated) hazard, the application of a control measure (standard/regulation) if applicable, to achieve Class 1 condition for the operator/ personnel where no further protective measures are required (other than proper training and testing for competency prior to use).

The NBHs are grouped first dealing with second-order optical radiation hazards, following with nominal equipment hazards.

Risk Assessment Worksheet										
NBH Control Measures – Part 5/5			Inherent Hazard/Risk				Residual Hazard/Risk			
These non-beam hazards are topical and for reference/guidance only. More comprehensive and applicable consensus standards and regulations will apply.		Controls Required	YES/NO/NA	Risk Index (Hazard Class)	Identify, if applicable, the appropriate control measure(s) taken to address the noted hazard.		Audit/Inspected	Pass/Fail/NA	Risk Index (Hazard Class)	Hazard Class
Z136.9 Clause	**Description**	**4**			**NBH Measure Solution**					**1**
7.2.2	Non-Laser Radiation (NLR): *Measures taken to address collateral, optical process, EM, RF, plasma, and/or ionizing radiation associated with LTIR or from the resonator operation.*	App		4	*Identify those NLR hazards generated/present with appropriate control measures taken.* *Ref.: ACGIH, NIOSH, ICNIRP, 29CFR1926.54, ANSI Z136.9-2013 (G4.2)*				1	✓
7.2.2.1.2	LTIR (UV & Blue-Light): *Macro laser materials processing interactions can generate optical re-radiation at the laser-target interaction zone. Potential for significant emissions in the 180-550 nm spectrum. High levels of UV can also produce significant amounts of ozone.*	App			*Ref.:* - *NIOSH 78–138* - *ANSI/IESNA RP-27* - *ANSI Z136.1-2007 (4.6.2.5.3)* - *ANSI Z136.9-2013 (G4.2.2)* - *ANSI Z87.1, IEC 62471*					
7.2.2.1.2	LTIR Optical/Process (VIS): *Macro laser materials processing interactions can generate optical re-radiation at the laser-target interaction zone. Potential for significant emissions in the visible (400-700 nm) spectrum, requiring appropriate attenuation measures for direct viewing.*	App			*Ref.:* - *ANSI/IESNA RP-27* - *IEC 62471* - *ANIS Z136.9-2013 (G.4.2.2)*					
	LTIR Thermal (NIR): *The thermal re-radiation of the laser-target interaction can present durability issues on nearby tooling and equipment, presenting a potential for failure.*	App			*E.g. laser cell tooling designed for and equipment specified to meet MIG welding best practices.*					
	LTIR Plasma: *High intensity interaction of the laser beam for materials processing can liberate material in the form of a vapor plume, in which further sustained interaction can elevate the plume temperature to a higher level generating plasma.*	App			*E.g. control measures to address the effects of plasma radiation to safeguard personnel.* *Ref. Z136.9-2013 (G1.2), (G4.2.1, G4.2.2)*					
	LTIR Ionizing (incl. Bremsstrahlung): *This is the next order of LTIR beyond plasma effects from higher intensity beams on the order of 10^{18} W·cm^{-2}.*	App			*E.g. Principles, objectives and requirements for ionizing radiation control measures found in 21CFR1020.40, 29CFR1910.1096* *Health effects: OSHA Directive TED 01-00-015 [TED 1-0.15A], (January 20, 1999).* *Ref. ANSI Z136.9-2013 (G4.2.1),*					
	Resonator, Collateral: *The (collateral) pump energy necessary for and/or the byproduct of resonator operations can generate incoherent optical radiation requiring appropriate control measures. These are grouped into the following domains:* - *Optical (e.g. pump diodes and arc/flash lamps)* - *Radio Frequency & Micro Wave (e.g. for excitation)* - *Power Frequency (Extremely Low Frequency) (e.g. with electrical excitation)* - *Plasma (such as with lasing gases)* - *Ionizing (High voltage (> 15 kV) power supplies can generate ionizing radiation)*	App			*E.g. nominally addressed through containment measures in the construct of compliant/certified laser (resonator). User manual should have appropriate inherent hazard information, respective control measures for safety awareness and maintenance, with identification of any residual hazard and additional PPE (or other control measures) if necessary for the operator.* - *Optical Ref.: ANSI/IESNA RP-27, IEC 62471* - *RF, ELF Ref.: IEEE C95.1, C95.6, C95.7* - *Plasma Ref.: ANSI Z136.9-2013 (G1.2, G4.2.1, G4.2.2)* *Ionizing Ref.: ANSI Z136.9-2013 (G4.2.1), OSHA Directive TED 01-00-015 [TED 1-0.15A], (January 20, 1999)*					

7.2.1 7.2.1.4	**Electrical:** *Electrical supply, distribution and control system to applicable safety standard.*	X	YES	4	*Ref.:* - *29 CFR 1910 Subpart S, 29 CFR 1910.147* - *NFPA 70 & 79, UL 508* - *IEC 60204-1* - *ANSI Z136.9-2013 (G4.1)*			1	✓
	Controls safety architecture: *Controls safety integrity/performance level commensurate with the hazard and work environment.*		App	4	*Ref.:* - *NFPA 79, UL 698* - *ANSI B11.TR6* - *EN954-1 (withdrawn)* - *IEC 61508, ISO 13849-1, IEC 62061*			1	✓
	Misc. electrical: *Ensuring electrical build conformance will enable safe use. Reasonably foreseeable fault conditions may include:*		App	4	*During design review(s), ensure that such topics are considered, assessed and addressed.* *Ref.:* - *ANSI B11.TR3, B11.0*				
	- Direct contact with live circuits, such as during maintenance and service - Indirect contact with circuits which have become active under faulty conditions - Approach to live circuits under high voltage (arc-flash) - Electrostatic phenomena - Resistive heating of circuit components under failure conditions								
7.2.5	**Mechanical, Associated with Robots and Automated Equipment:** *Automated equipment, including robots are to be rendered safe when in the presence of humans or have appropriate safeguards.*		App	4	*Normally addressed with level 'B' and 'C' standards.* *Ref.:* - *ANSI/RIA 15.06* - *ANSI B11.19, B11.20, B11.21, B11.25*				
	General & misc. mechanical: Ensuring electrical build conformance will enable safe use. Reasonably foreseeable fault conditions may include a variety of misfeeds or loss of sequencing in operations which can lead to a loss of hazard(s) containment or pose new hazards to personnel.				*Principles include containment, control and guarding measures as identified in:* - *29CFR1910.212 (General guarding requirements for all machines)* - *ANSI B11.0, B11.TR3*				
7.2.3	**Fire:** *Class 4 laser beams represent a fire hazard potential. Identify appropriate measures taken to protect against reasonably foreseeable operational or fault conditions in which a fire can be started.*		App	4	*E.g. laser cell tooling designed for and equipment specified to meet MIG welding best practices.* *Ref.:* - *NFPA 115, NFPA 51B* - *ANSI Z136.9-2013 (G1.1.2, G4.3)*			1	✓
7.2.4	**Explosion:** *Potential sources for explosion are from stored energy (e.g. capacitor banks) or*		App	4	*Ref.:* - *ANSI Z136.9-2013 (G1.1.3, G4.3)*				
	compressed gases (arc lamps, oxygen tanks/lines), particulate accumulation in fume extraction systems, LTIR failures.								
7.2.6	**Noise:** *Laser resonator, laser-target interaction (LTI) can generate excessive noise. In many cases, sound levels will not result in over-exposure, but may be a nuisance that should be addressed.*		App	4	*Ref.:* - *29CFR1910.95* - *ANSI Z136.9-2013 (G1.3)*				
7.3.1	**Laser Generated Airborne Contaminants (LGACs):** *Laser-target interaction may release various gases, fumes and/or particulates. Appropriate industrial hygiene safety control measures are to be applied.*		App	4	*Ref.:* - *ANSI Z136.9-2013 (G2.1, G4.4.1)* - *AWS F1.6, F3.2* - *ACGIH Industrial Ventilation Manual*				
7.3.2	**Compressed Gases:** *Lasing process, assist or shield gases in use for an application will require appropriate safety control measures for their delivery, use and evacuation.*		App	4	*Ref.:* - *29CFR1910.101, 29CFR1910.169* - *CGA P 1* - *ANSI Z136.9-2013 (G2.2, G4.4.2)*				

7.3.3	**Laser Dyes and Solvents:** *The organic dyes and solvents used in dye lasers are hazardous and may be toxic, carcinogenic and/or flammable.*	App	4	*Refer to respective SDS.* *Ref.:* - *ANSI Z136.9-2013 (G2.3, G4.4.3)* - *NFPA (30, 45)*
7.3.4	**Assist Gases:** *Safe delivery, deployment and evacuation of assist gases requires appropriate safeguarding. The assist gas can also influence the composition of LGACs as well as spectral response of the laser-target process radiation and NLR.*	App	4	*Ref.:* - *ANSI Z136.9-2013 (G2.2, G4.4.2)*
7.3.5, 7.3.5.1, 7.3.5.2	**Laser Process Area Ventilation, Prioritization:** *Local containment and control of LGACs is most effective when practicable.*	App	4	*Where utilized, exhaust ventilation recirculation is to comply with ACGIH Industrial Ventilation and Fundamentals Governing the Design and Operation of Local Exhaust Systems, ANSI Z9.2.*
7.3.6	**Process Isolation:** *A combination of process isolation separate from human access barriers may be required to provide a sufficient margin of safety.*	App	4	*Identify, if applicable, the levels or layers of process isolation required to achieve the necessary margin of safety and compliance.*
7.3.7	**Sensors and Alarms:** *Active monitoring of system critical safety elements may be required, commensurate with the hazard-risk, such as for hazardous gas cabinets, toxic or corrosive agents and gases, LGAC filters, etc.*	App	4	*Identification of those mission critical elements that are safety monitored, in a similar manner that safety interlocks are dual channel and monitored.*
7.4	**Biological Agents:** *For those lasers dealing with biological agents, etc., mapping of all process pathway and transport vectors with appropriate safeguards is required.*	App	4	*Additional regulations and standards will apply if dealing with biological substances/agents in manufacturing environments. Such will include but are not limited to: ANSI Z136.3* *Ref.:* - *ANSI Z136.9-2013 (G3, G4.5)*
7.5.1, 7.5.2, 7.5.3	**Ergonomics:** *Applicable human-machine interaction and interface is to be considered for safe operations. Noted human factor considerations include:* - *Limited Work Space* - *Work patterns* - *Operator postures* - *Consideration of hand-arm or foot-leg anthropometry* - *Sufficient local lighting for tasks* - *Mental overload and underload, stress* - *Human error, human behavior* - *Inadequate design, location or identification of manual controls* - *Inadequate design or location of visual display units*	App	4	*Ergonomic considerations improve the effectiveness that personnel can interact with equipment and minimize those contributing factors that can lead to an error in judgement regarding safety with the potential for an accident.* *Ref.: ANSI B11.TR1*
7.5.4.1	**Laser Disposal:** *Considerations for transfer/disposal of the laser system (a hazard source) have personnel and environmental safety obligations as well as possible restricted technology export constraints.*	App	4	*If the laser is being transferred, resold, or donated, it will be considered as "entering into commerce" and is subject to the requirements of all applicable product safety standards and regulations including 21CFR1040.10 and applicable export controls.* *Ref.:* - *ANSI Z136.9-2013 (7.5.4.1)* - *ANSI Z136.8-2012 (4.4.5, 4.7)* - *For product compliance:* - *21CFR1040.10 and 1040.11* - *21CFR1010.2 and 1010.3* - *21CFR1002.10* - *For export controls and license:*

				- The Department of Commerce, Export Administration Regulations (EAR – 15CFR 730-774), - The Department of State, International Traffic In Arms Regulations (ITAR – 22 CFR 120-130) - The Treasury Department, Office of Foreign Assets Control (OFAC –31 CFR 500599)			
7.5.4.2	Laser Waste Disposal: *Ensure conformance with regulatory (local, state, federal) requirements for hazardous laser waste (e.g. LGAC filters).*	App	4	The user manual for the laser system should have information regarding its safety life cycle for the owner/operator. The laser would be considered a hazardous waste (electronic plus associated lasing medium and operating fluids). *Disposal of the laser and its by-products (e.g. LGAC filters, coolant fluids) are subject to applicable local, state and federal regulations.* *If disposing of the laser, it may be necessary to dismantle it to allow for appropriate recycling.* *Ref.:* - *29CFR1910 Subpart H* - *29CFR1910 Subpart Z* - *EPA National Strategy for Electronic Stewardship (NSES)*			

LEGEND:

X Shall   — No requirement   **NHZ** Nominal Hazard Zone analysis required

● Should   **LSP** Laser Safety Program   ∇ Shall if enclosed Class 3B or Class 4

LSO Laser Safety Officer Determination

App Application dependent, LSO to determine if hazard potential exists

N.A. Not Applicable, addressed by other mutually exclusive criteria (e.g. 4.4.2.7.1 vs 4.4.2.7.2 vs 4.4.2.7.3)

N.R. Not Required, for the residual hazard class, addressed through integration efforts as noted

✓ See supporting documentation for conformance assessment and information

Chapter 6

Laser protective eyewear, looking sharp in the laser lab

Ken Barat

6.1 Introduction

I was once told that the use of laser protective eyewear is an admission of a laser safety failure. In fact with adequate beam enclosure and steps such as remote viewing the statement is true. Those controls will provide superior protection over eyewear.

But for many, when thinking of laser safety, eyewear is the first approach. There are numerous factors that go into eyewear. The most common approach is to answer these questions:

- What wavelength(s) am I trying to block?
- What is the minimum attenuation I need, meaning optical density?
- How well can I see with the eyewear on?

There are other fine detail considerations that affect eyewear and that the user may need to be aware of. But it is the questions above that drive most eyewear selection.

The ANSI standard gives the following as the list of items to be considered for full protection eyewear, that which will attenuate the incident beam to at or below the MPE:

- (a) Laser power and/or pulse energy.
- (b) Wavelength(s) of laser output.
- (c) Potential for multi-wavelength operation.
- (d) Radiant exposure or irradiance levels for which protection (worst case) is required.
- (e) Exposure time criteria.
- (f) MPE.
- (g) Optical density requirement of eyewear filters at laser output wavelength(s).
- (h) Angular dependence of protection afforded.
- (i) Visible luminous (light) transmission (VLT) requirement and assessment of the effect of the eyewear on the ability to perform tasks while wearing the

eyewear. From a practical standpoint, when the VLT is less than 20%, there may be insufficient light to perform the intended task.

(j) Need for side-shield protection and maximum peripheral vision requirement; side shields shall be considered and should be incorporated where appropriate.

(k) Radiant exposure or irradiance and the corresponding time factors at which laser safety filter characteristics degradation occurs, including saturable absorption especially for ultrashort aka ultrafast pulse lengths.

(l) Need for prescription glasses.

(m) Comfort and fit.

(n) Strength of materials (resistance to mechanical trauma and shock).

(o) Capability of the front surface to produce a hazardous specular reflection.

(p) Requirement for anti-fogging design or coatings.

As one can see it is a good sized list, each point has its value depending on the use situation. If one is using a carbon dioxide laser, most of the factors above lose importance. Why? Eyewear for CO_2 is plastic filter, lightweight and clear, as well as coming in many frame designs.

6.2 Eyewear labeling

Eyewear labeling is a simple item, legal laser eyewear must be labeled with optical density (OD) and wavelengths the filter is designed for. One should note that eyewear is almost always labeled with the most common wavelengths the manufacturer believes people are buying the filter for, not every wavelength it might attenuate. But, only what is labeled is what they are legally responsible for. Also there is no minimum font size for the labeling or one set location for the label to appear.

6.2.1 What does > or + mean?

Sometimes you will find a labeling that shows a range of wavelengths with a + or > near the OD, what does this mean? The answer is over that range the OD will never fall below that number. It does not mean the OD is sloping upward (figure 6.1).

6.3 Can eyewear break down/fail?

There are situations that strongly support the use of engineering controls over eyewear (figure 6.2). First, is recognizing that laser protective eyewear has limitations. Anyone who thinks they can stare down a 1000 W laser is in for a rude surprise. Glass filter, filters with dielectric coatings, and absorptive polycarbonate or other acrylic filters can only take so much irradiance and will be overcome in less than a second, certainly less than 5 s.

The impact of laser beams above some threshold will be converted to heat which will melt or crack the filter material.

For polycarbonate filters values are 10 J cm^{-2} ($t < 10^{-3}$ s) and 300 $t^{0.5}$ J cm^{-2} ($t \geqslant 10^{-3}$ s), where t is the exposure duration. For glass, the values are 1 J cm^{-2} ($t < 10^{-6}$ s) and 1000 $t^{0.5}$ J cm^{-2} ($t \geqslant 10^{-6}$ s).

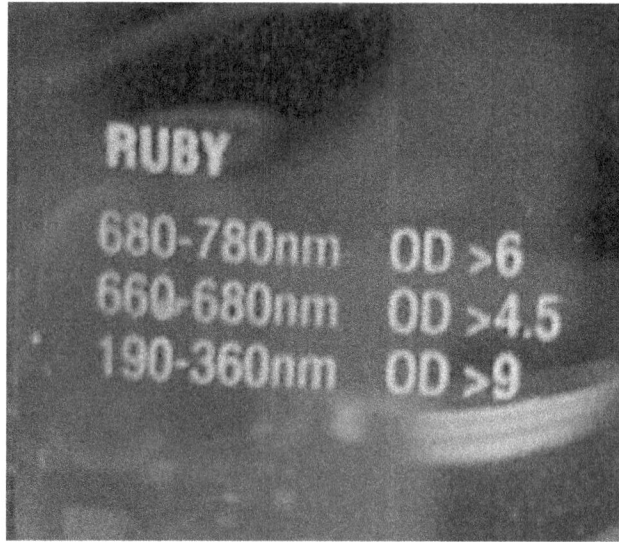

Figure 6.1. Use of greater than symbol.

Figure 6.2. Beam perpetuated eyeware.

Another consideration that leads to eyewear falling down on the job is, what is termed 'Saturable Absorption'. This is where the ability to absorb radiant energy decreases with increasing radiant exposure or peak irradiance. When this occurs, the optical density may decrease, providing less protection to the user. This has been reported for both glass and polycarbonate filters for certain pulsed lasers and is associated with high values of peak irradiance.

This is a well documented problem with pico-second and faster pulsed laser systems (femtosecond). Part of the problem is that the relaxation time of absorber materials is too slow between absorbing a photon, going to an excited state and then getting back to the ground state ready to absorb another photon. This has given rise to the 'M' rated eyewear.

If you are an LSO you can turn this into a *lessons learned notice*. Below is an example of a lessons learned notice that you can use to make your staff aware of this issue.

6.4 Ultrafast pulses and laser eyewear

At a glance

A study by the US Army Public Health Command demonstrated that the unique properties of ultrashort laser pulses influence laser protective eyewear filters. This effect can lower the optical density of the filters.

Details

Ultrashort pulses can cause shock-waves and bubble formation in the tissues of the eye, which lead to tissue damage.

An injury can occur from an intrabeam exposure or from viewing a reflection of the beam. Most documented injuries from ultrashort laser pulses have occurred when personnel view a reflection of the beam, while eyewear has been removed or one is looking under/over the eyewear. Exposure to just a single pulse, occurring on the order of a trillion times shorter than the blink of an eye, is all that is needed to cause damage.

The unique properties of ultrashort laser pulses make it difficult to protect personnel from them. The lack of protection, in this case from a lower optical density (OD) than that specified by a manufacturer for such short pulse durations can be contributed to by several factors:

- High peak power: Saturable absorption in the filter material may occur which reduces the OD.
- Absorptive material, does not have a quick enough relaxation time.
- Broadband emission: Supercontinuum generation converts a narrow wavelength band to a wide wavelength band which might require a broadband filter.

So how do I protect myself? Remote viewing is the second best option. Many common laser wavelengths, including those produced by the Ti:sapphire laser commonly used to create ultrashort pulses, can be viewed with inexpensive webcams or digital cameras. Another option is, when feasible, alignments should be made with the laser operating at a reduced energy or a longer pulse width.

The best option for protection during operation would be to enclose the beam path of the laser when emitting ultrashort pulses. This has the added benefit of protecting the optical components from dust.

What about M rated laser protective eyewear? M rated eyewear is narrow notch eyewear made to withstand ultrashort pulses. The problem is the few filters made this way are for very narrow wavelength bands.

So, should I wear my existing eyewear? The answer is YES, but awareness that possible reduction in OD is possible from a direct hit should make you even more careful to follow good practices.

Lessons learned completed, the material for this lessons learned comes from: US Army Public Health Command FACT SHEET 25-026-0614.

6.5 Angle of exposure

Are you exposed straight on or at an angle? Manufacturers of dielectric coating will tell you that the posted OD is only within a certain acceptance angle, similar to the acceptance angle of a fiber. Therefore, a beam entering at an angle outside the rated acceptance angle will be attenuated at a different level. The hope is that such beams will not be directed into the iris, but rather hit the white of one's eye.

6.6 Attacked from behind

This author only knows of two incidents where the user claimed to be struck by a laser beam from behind. In each case the beam entered from behind the person, and struck the inside of the eyewear and reflected back towards an eye. In either case the beam did not enter the pupil. Which goes back to the discussion about whether you can have gaps in your eyewear or should the eye area be completely blocked, such as with a wrap around or goggle style frame. In my mind this is really psychological to the users. I would rather have a pair I feel is comfortable to wear but has gaps than a pair that is uncomfortable without any. It is all a matter of how comfortable one feels about the concern. For me the likelihood of a beam entering and striking the inside lens and reflecting into a critical part of my eye is too low to outweigh wearing an uncomfortable frame.

6.7 Unusual eyewear event #1

A researcher recently received three pairs of laser safety eyewear manufactured by an established eyewear manufacturer, which contained filters that did not match what was stated on the product packaging or engraved on the filters themselves. The appropriate polycarbonate filters for the Nd:YAG/doubled Nd:YAG laser's wavelengths (1064 nm and 532 nm respectively) should have been amber or orange in color (as per catalog image), but the eyewear the researcher received had distinctively blue filters. The researcher noticed the issue when *he could see the beam* on the bench, which would not have been possible if the filters had been blocking the appropriate wavelengths. Metrologists responsible for transmittance calibrations confirmed by direct measurements that the filters received did not meet the specifications engraved on the filters. The researcher's institution contacted the manufacturer. The manufacturer confirmed the findings and has put a process in place to prevent the issue of mixing up filters.

What to take away from this

Be aware, check newly received eyewear over, speak up if things do not seem correct or what is expected. A Los Alamos researcher received a set of eyewear with a defect on the coating that allowed laser radiation through unfiltered. These things do happen, but let's not let it happen to you.

6.8 Unusual eyewear event #2

A worker, at a DOE Lab, was using a pair of dielectric coated (reflective) eyewear and viewed diffusely scattered 527 nm laser light through the filter media. This specific model utilized a 'notched' filter designed for OD7+@532 nm. The spectral data sheet from the manufacturer indicated the OD@527 nm is 3.1. The worker stopped work and borrowed a similar pair of glasses from a co-worker. They could no longer see the scattered 'green' light. The defective eyewear was taken out of service. No exposure to the worker was received from this event. A quick check, using a laser pointer, was performed and it was discovered that there was a problem with a couple of centimeters in the center of both lenses. A transmittance check was performed and it was found that at the center of each lens, the OD was only 1.5@532 nm. This event just recently occurred, so the investigation is still not complete. One big question to be answered is, was the dielectric coating indeed defective, or was it badly worn from improper use or storage? Stay tuned... One of the negatives mentioned concerning reflective eyewear is that it is angle sensitive. These filters are rated for an incident beam striking perpendicular to the media up to an angle of about 30° off axis. As the angle of incidence increases, the beam undergoes a shift in wavelength moving it away from the 'notched' protected wavelength. Simply put, you are not getting full protection and laser light may pass through the filter into your eye. Also, if using a reflective 'notched' filter that is rated for 532 nm, it may not provide protection for a 527 nm laser. Verify applicability either with the manufacturer or your Laser Safety Officer.

6.9 Absorptive versus reflective filter

Most of the laser eyewear on the market can be divided into absorptive or reflective when talking about how they deal with incident laser beams. Each has advantages and disadvantages.

The absorptive method is the most commonly used filter used in laser eyewear, chiefly due to cost and its light weight, making it easier to wear for prolonged periods. The positives are:
1. Inexpensive.
2. Can cover a large range of wavelengths.
3. Glass filters for IR protection are relatively clear.
4. Polycarbonate filters are lightweight and can be shaped into many styles.
5. Polycarbonate provides excellent impact resistance, which is not a requirement of laser eyewear.

On the flip side:
1. Glass filters are heavy compared to plastic lenses.
2. Limited in shape, therefore, limited frame designs.
3. Rarely impact resistant.

6.10 Impact resistance

Impact resistance for safety glasses is almost a default item, but not laser eyewear. The ANSI Z136.1 standard safe use of lasers does not require laser protective eyewear to be

ANSI Z87 compliant. ANSI Z87 is the standard for safety eyewear; the most common element is impact resistance. While evaluating your eyewear choices—the question of impact resistance may come up. Simply: is it needed or not? If not then no further action is needed; if impact resistance is needed there are three choices:

1. Obtain a pair of laser eyewear that is compliant with Z87.1 (most polymer eyewear is compliant).
2. Wear safety glasses over the laser eyewear, rather poor option.
3. Have glass laser eyewear hardened to meet Z87.1.

Choice 2 can affect comfort or the ease of wearing the protective eyewear and general vision, while choice 3 will affect the cost of the eyewear. Glass eyewear is not impact resistant unless hardened (tempered/laminated).

6.11 Manufacturer protection curves and non-labelled wavelengths

Please remember while manufacturers, either on their website or at your request, have curves for each of their filters showing OD for different wavelengths for a particular filter, they are only legally responsible for the wavelength and OD that is printed on the eyewear. If you wish to use wavelengths outside what is labeled, ask the manufacturer to relabel or for official documentation of protection.

6.12 What to do if labeling wears off?

There is a good chance that over time the labeling on ones eyewear will wear off. This is particularly true if the labeling is printed on the eyewear. So what do you do when that happens? The first thing you need to know is, the eyewear is illegal to use without labeling. But do not throw it away. You can relabel the eyewear as long as you know the wavelength and OD. Another approach is to place a unique identifier on the eyewear and post a chart describing the eyewear and the wavelength coverage and OD, see below as an example (figures 6.3 and 6.4).

6.13 Prescription eyewear

Ground filter lenses are very rare, the answer is laser eyewear with a prescription insert (figure 6.5). Which gives the wearer a custom pair of eyewear. Inserts can be removed and new lenses put in as the user's prescription changes.

6.14 Alignment eyewear

A common reason for not wearing laser eyewear is 'I cannot see the beam to align the system'. To help answer that question one should really use remote viewing, but since most folks do not think of that or believe the cost will be too much, alignment eyewear has come about. Such eyewear should only be used for visible beams. The goal is that one will be able to see the beam but slightly dimmer. Most importantly, if one is struck by the beam, the beam transmitted through the filter will be the intensity of a Class 2 or no more than a Class 3R. Bright enough for one aversion response to kick in, but not intense enough to cause injury. EN 208 is a standard

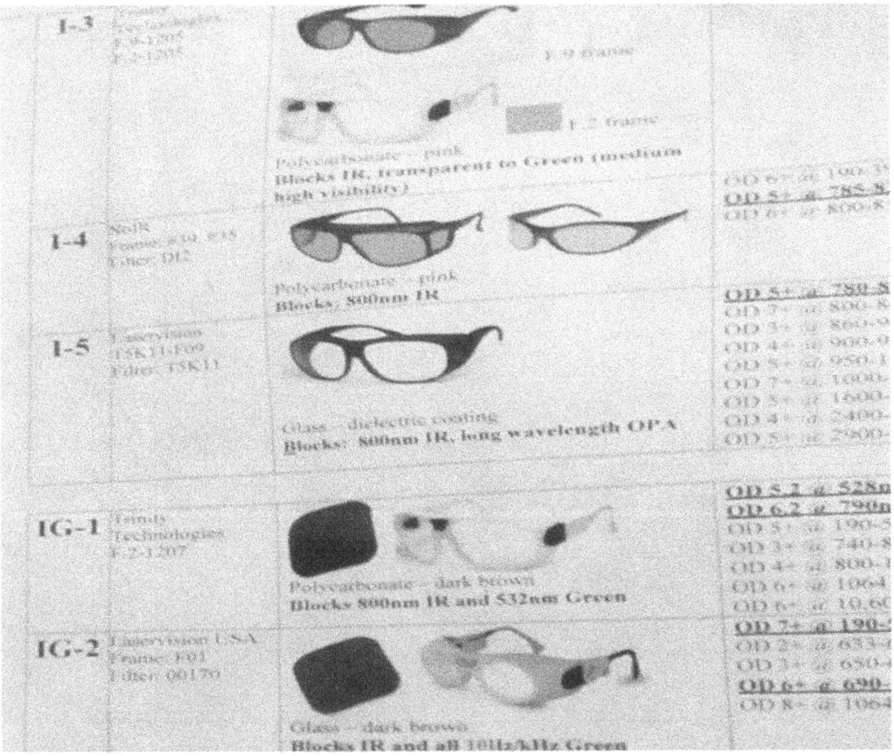

Figure 6.3. Eyewear chart.

which only deals with alignment eyewear choice. An existing European norm which recommends optical density for alignment eyewear versus the output of lasers used.

Scale	OD	Maximum instantaneous power continuous wave laser (W)	Maximum energy for pulsed lasers (J)
R1	1–2	0.01	2×10^{-6}
R2	2–3	0.1	2×10^{-5}
R3	3–4	1.0	2×10^{-4}
R4	4–5	10	2×10^{-3}
R5	5–6	100	2×10^{-2}

Therefore, for 'alignment' laser eyewear to be effectively utilized, preferentially all of the following conditions should be in place: administrative liability acknowledgment and acceptance of same, acknowledgment of potential hazards with the utilization of eyewear that does not protect one against small source intrabeam or specularly reflected exposures and finally, collaborative agreement between the LSO and researcher(s) of alignment eyewear safety protocols and appropriate 'alignment' laser protective eyewear. Once these preliminary 'philosophical' protocols are established, the implementation of alignment eyewear can proceed.

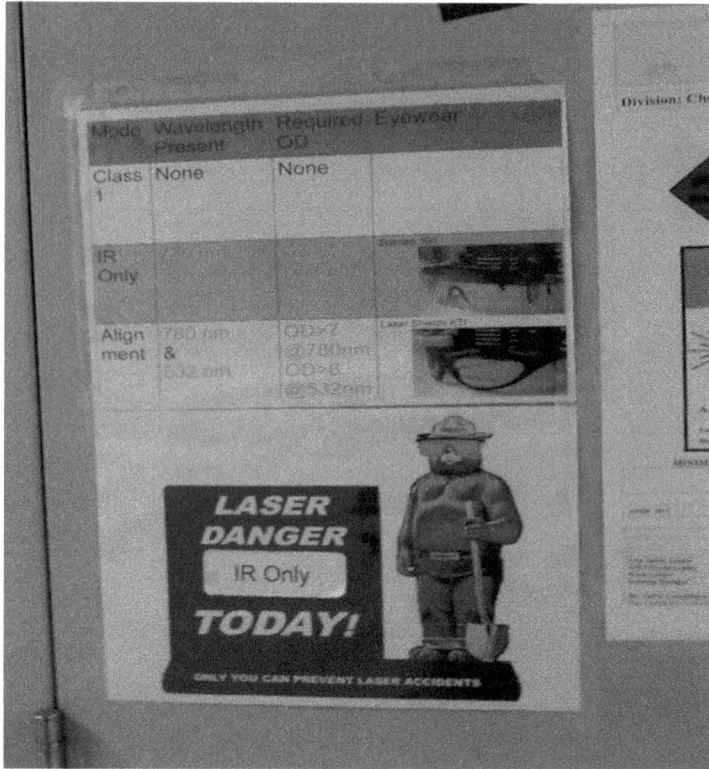

Figure 6.4. Getting the message across.

Figure 6.5. Insert for prescription.

The trouble with EN207 is on the pulse side, it does not cover today's laser pulses nano-, pico- and femtoseconds. My experience has been a decrease of 1.4 OD is the maximum for alignment eyewear used for pulsed lasers.

6.15 European labeling

If you look at the labeling of your eyewear and it is similar to what you see in (figure 6.6), that is the European system. This, while containing more information for the user than the US ANSI system, is not obvious. One must know the code to make any sense of it or interpret the code. In addition the eyewear must have a CE marking.

Protection level

690–795 DIRM LB7 (OD 7+).
>795–805 DIR LB7 + M LB9Y (OD 9+).
>805–1100 DIR LB7 + M LB9 (OD 9+).
>1100–1200 DIR LB7 + M LB8 (OD 8+).
>1200–1320 DIRM LB7 (OD 7+).
>1320–1400 DIRM LB3 (OD 3+).
1400–1550 DIR LB3 (OD 3+).
10 600 DI LB4 (OD 4+).
633 0,1@ 2 × 10E−5J RB2 (OD 2–3).

EN code

- **D** Continuous wave CW ('Dauerstrich') 10 s.
- **I** Normal pulse ('Impuls') ~ ms > 1 μs and <250 ms.
- **R** Q-switched ('Riesenpuls') ~ <1 μs.
- **M** Mode-locked < ns ~ 'Femtosecond'.
- **L** is the scale number, similar to OD, but with an irradiance protection factor. A letter after the L, such as LB indicates how current the scale number is, and which standard version it was developed to.

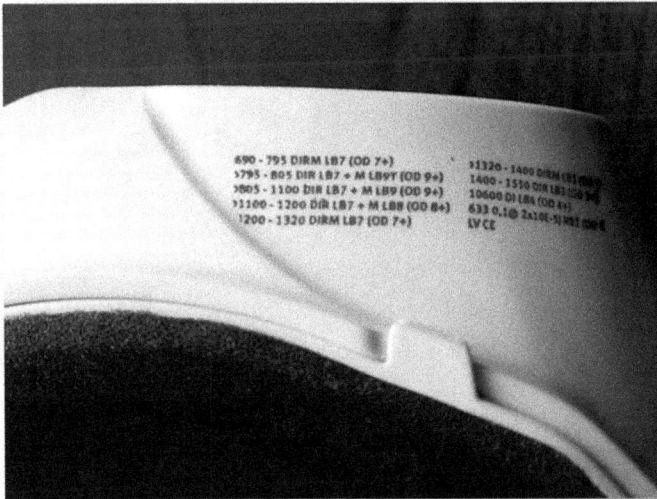

Figure 6.6. European labeling.

6.16 Storage of eyewear

Your eyewear cannot take care of you if you do not take care of it. How one stores it is important to its useful lifetime (figures 6.7 and 6.8). Laying your eyewear down on the lens is never a good idea. Hanging eyewear by elastic straps is definitely a NO-NO (figure 6.9). Please take care of your eyewear, while expensive it can last for years if treated right, just like a good marriage (figure 6.10).

Figure 6.7. Appropriate storage.

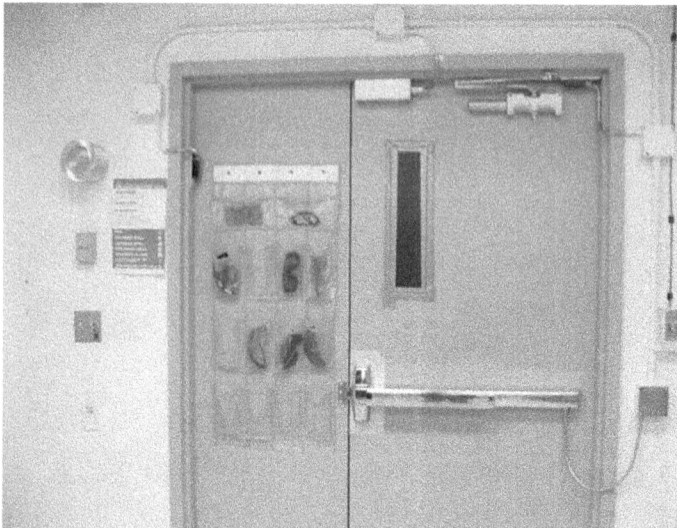

Figure 6.8. Another good solution is using a shoe tree for eyewear storage.

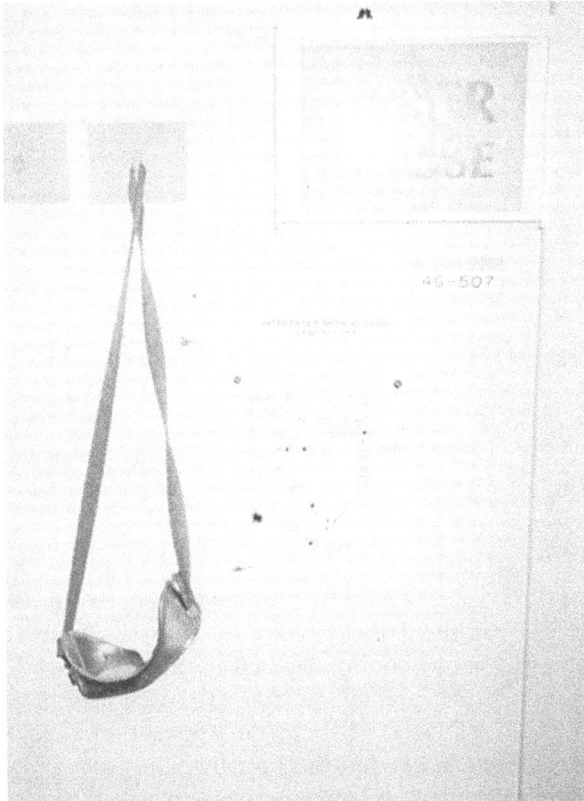

Figure 6.9. Elastic straps stretched out due to being hung.

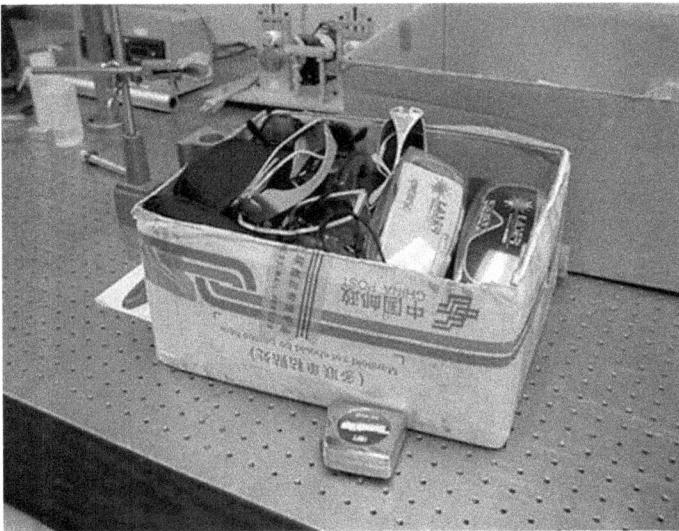

Figure 6.10. Random box storage, wrong technique.

Chapter 7

Regulations, you mean there are rules?

Ken Barat

7.1 Introduction

While laser use has spread across all sectors of modern society, it has not escaped the attention of standards and regulatory bodies. Of course, whenever regulations are applied to a technology, that technology often leaps ahead and with the slow pace of regulatory change, regulations are always trying to catch up and be relevant. Rules for many laser applications lag far behind the application. This is a problem for those using and developing lasers and their application.

As an example, consider the Class 4 argon laser, a classical laser. One to which the majority of laser rules, whether manufacturing or lab use, easily apply. Now consider the Class 4 laser diode, fiber laser or diode array. While it is easy to interface an argon laser to a room interlock or emergency shut-off button, the same cannot be said for a laser diode array on a breadboard that can easily be moved, not only around a room, but from room to room.

7.2 Standards and regulations

Standards and regulations fall into two major groups. The first is product safety. What elements are needed or considered needed for a safe product? These include engineering controls, such as protective housing interlocks and emission indicators to administrative controls. This commonly includes labeling and user manuals. To digress for a minute, once again consider the argon laser; it is easy to label, however, in most diode cases, the label is larger than the diode itself.

Once we have made a safe product, user guidance and regulations come into play. The majority of laser user regulations come from the adoption of user standards by regulatory bodies. This applies to the United States and the American National Standards Institute (ANSI) Z136 series on safe use of lasers and the rest of the world following International Electrochemical Commission (IEC) 60825 series. In both cases, regulatory bodies can and do make modifications and additions to suit their own perspective. One easy example is the reporting of laser injuries to regulatory

doi:10.1088/2053-2563/ab0f25ch7

7-1

bodies. It is common for National Regulation and State Agencies to require anything from immediate or 2 h–48 h notification of a laser injury, particularly when it applies to an eye injury.

There are three regulatory groups often overlooked when one talks about laser regulations. They are: outdoor, outer space rules and export restrictions. While many nations have rules and restrictions on what they consider sensitive technologies and try to keep within their borders, the example I will use is the United States. Then we will be back to the routine regulatory groups.

7.3 Export control

Every LSO needs to beware of limitations on laser products and technology rules for their nation. Because these limitations not only apply to the hardware but can extend to information including any oral, written, electronic or visual disclosure, shipment, transfer or transmission of commodities, technology, information, technical data, assistance or software codes. Wow, that covers a lot. The reason most commonly given is national security or protection of trade. For history buffs the same applied to chocolate and tulips.

7.3.1 University guidance examples

If the reader thinks universities are exempt or do not care about this, here is official guidance from two universities, the first is the position of the University of Kansas to its population:

'The term "export controls" typically refers to regulations overseen by several federal agencies, especially the Departments of State, Commerce, and Treasury, that implement federal laws put in place to protect national security, promote foreign policy, and in some cases to control short supplies. The regulations govern the international transfer of military and most commercial items, including software and technical information, and certain services. In addition to the transfer of items out of the United States, the term "export" also refers to the release of controlled software source code or technical information to a foreign national, whether in the US or abroad.

It is important for all members of the University community to understand how export control regulations can affect their University activities, identify export control issues, and obtain assistance to properly document exports.

Export control regulations are intricate and involved. Penalties and fines imposed for export control violations are severe and can include criminal, civil and administrative charges. You are encouraged to seek assistance for any question on the topic.

The University Office of Export Compliance (OEC) is responsible for directing and monitoring the University's export control compliance program and implementing procedures and/or guidelines to comply with federal export control laws and regulations. For more information visit the OEC website or contact the KU Export Control Officer at ueco@ku.edu.'

7.3.2 Example #2

Second, here is policy from Ohio State University, which provides even more detailed instructions:

What do OSU personnel need to do?

In order to ensure compliance with export controls, it is critically important for university personnel to identify when their activities may trigger export controls. When export controls apply, individuals must take the appropriate steps to obtain any required governmental licenses, monitor and control access to restricted information, and safeguard all controlled materials.

What kinds of activities might trigger export control issues?

Research in export restricted science and engineering areas—examples include:
- Military or defense articles and services.
- High performance computing.
- Dual use technologies (technologies with both a military and commercial application).
- Encryption technology.
- Missiles and missile technology.
- Chemical/biological weapons.
- Nuclear technology.
- Select agents and toxins (see Select Agent/Toxin list).
- Space technology and satellites.
- Medical lasers.

Traveling overseas with high tech equipment, confidential, unpublished, or proprietary information or data

Traveling with certain types of high tech equipment including, but not limited to, advanced GPS units, scientific equipment, or with controlled, proprietary or unpublished data in any format may require an export license depending on your travel destination. See International Travel for more information.

Traveling with laptop computers, web-enabled cell phones and other personal equipment

Laptop computers, web-enabled cell phones, and other electronics containing encryption hardware or software and/or proprietary software can require an export license to certain destinations. In general, an export license will be required to take any items to or through any US sanctioned country (e.g. Iran, Syria, Cuba, Sudan, and North Korea).

Use of 3rd party export controlled technology or information

University activities involving the use of export controlled information, items, or technology received from outside the university are not protected under the Fundamental Research Exclusion and all research involving the use of export restricted technology is subject to all export controls. For help in determining export control issues see Incoming Export Control Information Questionnaire.

Sponsored research containing contractual restrictions on publication or dissemination

The vast majority of research done at the university is shielded from export controls under the Fundamental Research Exclusion. However, this protection is lost whenever the university or the researcher agrees to allow any restrictions on the publication, dissemination, or access to the research by foreign nationals.

Shipping or taking items overseas

University activities that involve the transfer of project information, equipment, materials, or technology out of the US by whatever means will be subject to export controls and may require export license(s) depending on the item, destination, recipient, and end-use.

Providing financial support/international financial transactions

University activities that involve the international payment of funds to non-US persons abroad need to be verified to ensure that the university is not inadvertently providing financial assistance to a blocked or sanctioned entity. Examples include providing support via a subcontract to a non-US university or providing payments to research subjects in other countries. Contact exportcontrol@osu.edu if your activity involves payment to persons or organizations outside the US.

International collaborations and presentations

University activities that involve foreign national faculty, students, staff, visiting foreign scientists or collaborator(s), or other foreign entities (e.g. non-US company, university or other organization) or research that will include travel to international conferences to present unpublished results may be subject to export controls especially if any of the foreign nationals are from embargoed or sanctioned countries. See International Collaborations for more information.

International field work

Research projects where any part of the research will take place outside the US (e.g. field work outside the US) may not qualify under the Fundamental Research Exclusion and may be subject to export controls. For help in determining potential export control issues see the International Research Export Control Questionnaire.

International consulting

Providing professional consulting services overseas, especially to embargoed or sanctioned countries (e.g. Iran, Syria, Cuba, Sudan and North Korea) is, in most cases, strictly prohibited.

Where can you get help?

This website has been designed to help you understand and comply with the export control regulations. Information on various export control topics can be found by clicking on links above or in the right-hand menu bar. Assistance can be obtained by contacting the Export Control office at exportcontrol@osu. eduor by contacting the Office of Research Compliance at (614) 292-4284.

All I can say is, if your institution is not aware of export control, someone needs to see that it is addressed. The use agencies most involved are the following:
   a. Export Administration Regulations (EAR—**15 CFR** 1264 **730–774**), **The Department of Commerce's**, which covers the export of goods and services identified on the Commodity Control List.
   b. International Traffic in Arms Regulations (ITAR—**22 CFR** 1267 **120–130**), **The Department of State's** also known as the US Munitions List, which covers defense-related items and services.
   c. The **Treasury Department's** Office of Foreign Assets Control (OFAC—31 CFR 500–1270 599), which covers control of certain products to certain countries.

7.4 Outdoor use

Let us turn our gaze up to the heavens. Any laser user that may impact navigable airspace needs to be aware that here are regulatory agencies and standards that need to be followed and notified of the activity. There is the obvious concern of distracting pilots, then you need to consider whether you are impacting a satellite in space. Soon it will be being a threat or nuisance to a drone, delivering someone's groceries or worse.

Pilot illumination is a major concern to the air traffic community. The source, at least so far, is to general population users of legal laser pointers (5 mW output limit) and users of higher power hand held lasers; who do not realize or recognize the threat they pose.

Just be aware that more and more countries are making it a major crime to illuminate objects in the sky. While actual criminal cases taken to court are few compared to the number of reports, they do happen.

7.4.1 Non-MPE effects

All I want to say on this topic is that visible laser beams can cause effects that interfere with the drive or pilot function but are below the threshold to cause permanent eye injury. Glare, distraction and flash blindness are the terms used to describe these effects at different power or exposure levels. All these effects are commonly discussed when talking about pilot illumination by hand held lasers. This is a worldwide problem and people are trying to address it more through education than regulations. A number of regulations exist worldwide, but it is the education efforts that will have the greatest effect to stop or reduce this problem.

7.5 What if you manufacture lasers?

To legally sell a laser within the United States, and the majority of nations, some product safety rules apply. In the US, no matter where it was manufactured a laser must comply with the Federal Laser Product Safety Standard, which are regulatory rules not a standard. It is a branch of the US Food and Drug Administration that sets the safety items for laser products. That branch is the Center for Radiological Devices and Health (CDRH). CDRH propagates laser safety product rules in 21 CFR1040 (available for download from the CDRH web site). Laser products are required to be made and certified to their regulations, and to carry a certification label. It is still legal to sell uncertified lasers if the manufacturer has registered with CDRH and the laser is not made to be used in a standalone fashion, such as a diode laser that goes into a printer. Once again, any laser sold in the United States must be certified with the exemption cited. Certified lasers must have:
- Certification label.
- Identification label.
- Name and address of manufacturer.
- Place, month, and year of manufacture.
- Hazard classification.
- Radiation output info and warning logotype.
- Aperture label.

A common question concerning lasers or laser systems made or modified for internal institutional use is, 'Am I a laser manufacturer?' The FDA will not consider multiple products to have been 'manufactured' provided they are:
- Not shipped in interstate commerce.
- Used solely at the place where constructed.
- Used by the same employees who constructed them.
- Not made on a recurring basis.

Laser products for introduction into commerce in, or imported into, the United States must:

- Comply with 21 CFR sections 1040.10 and 1040.11 as applicable.
- Be certified and identified in accordance with 21 CFR sections 1010.2 and 1010.3.
- Be reported in accordance with 21 CFR section1002.10.

7.5.1 Outside the United States, meaning the rest of the world

In today's global world, life does exist outside the US. For lasers manufactured and sold outside the US it is compliance with 60825-1 that is the standard to follow. While 60825-1 and CDRH rules are close, there are a few differences. This gap has been narrowed by CDRH's Notice 56, which is an attempt by the CDRH to harmonize its rules with the rest of the world. Therefore, as an example, only one set of labeling is required on laser products. Until the CDRH formalized this harmonization in its regulations (CFR), a new Notice has to be issued each time 60825-1 is reissued with changes.

CDRH has issued notices to laser product manufacturers and importers stating non-objection to:

- Lack of emission indicators or beam attenuators on Class II and Class IIIa systems.
- Hazard warning labels as specified in IEC 60825-1.

CDRH will not object to conformance with many sections of IEC 60825-1, as amended, and IEC 60601-2-22 as alternatives to comparable sections of 21 CFR sections 1040.10 and 1040.11. CDRH plans to amend federal regulations for laser products to reflect those sections of the IEC standards. CDRH is also listing sections of its standard that contain requirements to which manufacturers must conform. This action is appropriate because of the Center's intent to harmonize its requirements with many of those of the IEC standards.

Effective immediately, and until the effective date(s) of any amendments of the Federal regulations affecting laser products, CDRH will not object to conformance with the comparable sections of IEC 60825-1, Editions 1.2 or 2 or 60-601-2-22 in lieu of conformance with the following sections of 21 CFR section 1040:

1040.10(b) Definitions
1040.10(c)(1) Classification
1040.10(d) Accessible emission limits
1040.10(e) Tests for determination of compliance
1040.10(f)(1) Protective housing
1040.10(f)(2) Safety interlocks
1040.10(f)(3) Remote interlock connector
1040.10(f)(4) Key control
1040.10(f)(5) Laser radiation emission indicator
1040.10(f)(6) Beam attenuator
1040.10(f)(7) Location of controls

1040.10(f)(8) Viewing optics

1040.10(f)(9) Scanning safeguard

1040.10(g) Labeling requirements

1040.10(h)(1) User information

1040.11(a) Medical laser products

CDRH intends to harmonize the requirements of these sections with those of the IEC standards.

What CDRH requirements or sections have not been affected by the harmonization effect? They are either beyond the scope of the IEC standard, are sufficiently different, or are not normative and included as recommendations in the User's Guide section of the IEC standard.

Laser products must conform to the following sections of the CDRH standards:

1010.2 Certification

1010.3 Identification

1010.4 Variances

1040.10(a) Applicability

1040.10(c)(2) Removable laser systems

1040.10(f)(10) Manual reset mechanism

1040.10(h)(2) Purchasing and servicing information

1040.10(i) Modification of a certified product

1040.11(b) Surveying, leveling and alignment laser products

1040.11(c) Demonstration laser products

7.6 Laser users

7.6.1 Outside the United States

The International Electrotechnical Commission (IEC) is a worldwide organization for standardization comprising all national electrotechnical committees (IEC National Committees). The object of IEC is to promote international co-operation on all questions concerning standardization in the electrical and electronic fields. IEC publishes International Standards, Technical Specifications, and Technical Reports.

60825-14 is the IEC document that outlines laser user conduct. Once the IEC standard is voted upon and approved by the relevant committee, it becomes the responsibility of each member nation to adapt the standard as its National regulations. In doing so each nation can include national specific items. This may include age limits to work with lasers or accident reporting requirements to National regulatory agencies.

The most well-known IEC laser standard is 60825-1, which is a combination of CDRH rules and ANSI Z136 controls. Outside of the United States, users follow IEC standards and regulations, yet within the US ANSI Z136 is the dominant user standard. The major IEC laser standards are:

- IEC 60825-1, for the Safety of Laser Products.
- IEC 60825-2, for the Safety of Fiber Optic Systems.

- IEC 60825-14, the laser user standard.
- IEC 60825-5 Manufacturers checklist.
- Many other IEC 60825 standards exist.

7.6.2 Within the United States

7.6.2.1 User safety

Laser users fall under the Occupational Safety & Health Administration, OSHA, which states simply that users comply with the American National Standards Institute Laser standard Z136.1 Safe Use of Lasers.

7.6.2.2 Lasers and OSHA

In a search of the OSHA regulations one will find very little on laser use. So how does OSHA extend its reach to the large community of laser users? By the adaption of 'National Consensus Standards'.

Section 6(a) of the Williams-Steiger Occupational Safety and Health Act of 1970 (84 Stat. 1593) provides that 'without regard to chapter 5 of title 5, United States Code, or to the other subsections of this section, the Secretary shall, as soon as practicable during the period beginning with the effective date of this act and ending 2 years after such date, by rule promulgate as an occupational safety or health standard any national consensus standard, and any established Federal standard, unless he determines that the promulgation of such a standard would not result in improved safety or health for specifically designated employees.' The legislative purpose of this provision is to establish, as rapidly as possible and without regard to the rule-making provisions of the Administrative Procedure Act, standards with which industries are generally familiar, and on whose adoption interested and affected persons have already had an opportunity to express their views. Such standards are either (1) national consensus standards on whose adoption affected persons have reached substantial agreement, or (2) Federal standards already established by Federal statutes or regulations.

The national standard OSHA looks to for laser safety compliance is Z136.1 Safe Use of Lasers. Every LSO needs to know that under Z136.1 section 1 they can use control measures from other Z136 series laser standards that are more relevant to their use application, even if it contradicts Z136.1.

OSHA also has a database of laser accidents, but it is underused and the majority of accidents reported to it are more non-beam than laser related. Here is part of that database:

Event date	File #	Event description
12/18/2001	50552652	Employee is overcome and killed by carbon monoxide

On December 18, 2015, an employee who worked on a hog farm entered a building that was approximately 14 feet × 17 feet with an 8 foot ceiling. It was divided into a primary entrance and an office. Employees are required to shower and change clothes prior to entering the hog barn. The employee reached through a window into the office to set down her bag and lunch, and entered the door to the shower. Her coworker, the owner, was in the hog barn cleaning the facility with an Alladin 16535 power washer that was installed in the office area near the shower. While on a break, the coworker went looking for the employee and found her unconscious in the shower area. After testing the employee's blood, it was found that the employee's carboxyhemoglobin level was at 47.8 percent per SLCLs calculation, the employee's exposure for a 60 min occupational exposure was 2772 ppm. The 8 h time weighted exposure was 346 ppm. The employee was killed by asphyxiation.

03/12/2012	0454510	Employee finger is amputated in Notcher machine
09/24/2011	0950633 3441	Worker sustains laceration when caught by machine
08/20/2010	0950643 7538	Employee is burned by hot water
06/07/2010	**0524530x3499**	**Employee dies in laser cutter accident**

On June 7, 2010 Employee #1 was operating an automated laser cutter within the production area. The front guard was raised and Employee #1 was making adjustments inside the point or operation area. Employee #1 became caught in a pinch point when the structure that holds the laser cutter moved to the right front area near the metal structure of the machine. Employee #1's head became caught in the limited space between the metal structures, approximately three inches wide. Another coworker heard a noise and found Employee #1 in the machine. The injury was immediately fatal.

06/01/2006	**10541115051**	**Employee's arm is amputated in laser cutter**

At approximately 1:45 p.m. on June 1, 2006, Employee #1 was removing a piece of metal from a laser cutting table. He reached over the table as the cutter proceeded through its cycle. The laser cutting head amputated Employee #1's left arm. He was hospitalized for treatment.

03/14/2003	**0950614 8221**	**Employee's eye injured by exposure to laser radiation**

On March 14, 2003, an employee of U C Berkeley, was working with a class 4 Nd: YAG near infrared open beam laser. He wore laser safety eyewear during beam alignment, but afterwards he removed safety eyewear. While taking a meter measurement for output energy, his eye caught a flash from a hidden optic in the beam's path. His exposure exceeded current limits for lasers. The exposure to laser damaged the macula in his retina, resulting in irreversible retinal damage and loss in visual acuity.

01/07/2003	0950644 3556	Employee's legs crushed while in laser cutting pit
08/26/1999	0950643 3444	Electric shock direct contact with energized parts
03/18/1996	**0950632 8221**	**Two employees suffer eye injuries from non-direct laser light**

At approximately 3:00 p.m. on March 18, 1996, Employee #1, an assistant project scientist working in the laboratory of Dr Kent Wilson at the University of California at San Diego, was aligning a 50 Hz, 800 nm tera-watt laser system for an experiment. As he was viewing the laser focus through a chamber window, he became exposed to nondirect laser light. The approximate laser intensity was 3 W at 50 Hz, producing about 60 mJ/pulse of energy. During this brief period of alignment, Employee #1 had removed his protective eyewear. He suffered serious injury to both eyes that involved partial vision loss. After this incident, it was determined by the U.C.S.D. Environmental Health and Safety Department that two similar incidences had occurred earlier to another worker. Sometime on the evening of December 28, 1995, Employee #2, a research chemist in the same lab and a coworker of Employee #1, was performing the identical procedure as previously described. During the period of alignment, he removed his protective eyewear and suffered serious injury to both eyes involving partial vision loss. Employee #2 sought medical attention at a later date. The second incident occurred on March 16, 1996. Again, alignment of the laser into the chamber was being performed. When Employee #2 removed his safety glasses to view the focus, both of his eyes were again exposed to nondirect laser light. The intensity and power of the laser were approximately the same as noted for Employee #1. The causal factors for all three incidences were unsafe work practices and, more specifically, removing protective eyewear during the alignment of the 50 Hz, 800 nm tera-watt laser system. The accidents occurred at an academic research laboratory that specialized in laser science and, specifically, in the construction of high intensity, ultra-short laser and x-ray systems to view molecular dynamics.

07/15/1991	0950623 8062	Employee's vision marred by dark spot
03/15/1989	03524220 2311	Employee's arm burned by carbon dioxide laser beam

7.7 US States that have a laser regulatory programs

A small number of US States have rules on laser use. They either center on medical lasers or Class 3B and Class 4 lasers. In addition, some states and cities have restriction on hand held lasers. These later rules apply to the illumination of aircraft. Some states have separate rules affecting medical lasers and who can use them (i.e. Physician versus Cosmetologist). Overall all the State programs go through swings of activity and times when they seem asleep at the wheel.

Alaska: Radiological Health Program section of State Laboratories Department of Health and Social Services: Title 18 of the Alaska Annotated Code Part 85, Art. 7, Sect. 670–730. (Oct. 1971 and Apr. 1973).

Arizona: Arizona Radiation Regulation Agency Regulations, now part of the Dept of Health Services: Article 14, Rules for Control of NIR, Sect. R12-1-1421 to 1444.

Florida: Florida Department of Health, Bureau of Radiation Control Regulations: extensive regulations in chapter 10D-89 of FL code.

Georgia: Office of Regulatory Service Department of Human Resources Regulations: registration requirements in chapter 270-5-27, GA Code (9/1/71).

Illinois: Division of Electronic Products Department of Nuclear Safety Registration Regulations in Laser Systems Act of 1997 (effective 7/25/97).

Massachusetts: Massachusetts Radiation Control Program Regulations: Registration and Control regulations (ANSI Z136 based). Effective 5/7/97.

New York: Department of Labor Radiological Health Unit Regulations: In Industrial Code Rule 50 of Title 12 (cited 12 NYCRR Part 50). Amended 3/2/94.

Texas: Bureau of Radiation Control Department of Health Division of Licensing, Registration and Standards. Texas Regulations for the Control of Laser Radiation Hazards (TRCLRH), Parts 50, 60, and 70. 25 Texas Administrative Code (TAC) 289.301.

7.8 ANSI Z136

While the American National Standards Institute (ANSI) Z136 laser series is designed only as guidance for laser users, it has taken near regulatory impact since OSHA, US Federal Agencies and several State regulatory programs audit against it.

The ANSI Z136 laser series is broken down into a Horizontal Standard Z136.1 Safe Use of Lasers which cuts across all laser application standards and a number of application specific standards (termed vertical standards).

The rest of the Z136 Laser Series is made up of 'Vertical or Application' standards.

- ANSI Z136.1—*for Safe Use of Lasers.*
- ANSI Z136.2—*for Safe Use of Optical Fiber Communication Systems Utilizing Laser Diode and LED Sources.*
- ANSI Z136.3—*for Safe Use of Lasers in Health Care.*
- ANSI Z136.4—RP *for Laser Safety Measurements for Hazard Evaluation.*
- ANSI Z136.5—*for Safe Use of Lasers in Educational Institutions.*
- ANSI Z136.6—*for Safe Use of Lasers Outdoors.*
- ANSI Z136.7—*for Testing and Labeling of Laser Protective Equipment.*
- ANSI Z136.8—*for Safe Use of Lasers in Research, Development or Testing.*
- ANSI Z136.9—*Safe Use of Lasers in Manufacturing Environments.*
- Draft Z136.10—*Safe Use of Lasers in Entertainment, Displays and Exhibitions.*

7.8.1 Special note overlooked by many

Section 1 of Z136.1 state that control measures in the vertical standard can take precedence over sections in Z136.1 if the LSO deems them to be more appropriate for the laser applications under their review; even if they contradict Z136.1. Therefore, the LSO can choose controls from more than one standard in building their laser safety program.

Wording found in the ANSI standard is generally some version of the following:

The objective of this standard is to provide reasonable and adequate guidance for the safe use of lasers and laser systems. A practical means for accomplishing this is first to classify lasers and laser systems according to their relative hazards and then to specify appropriate controls for each classification.

Other special application standards within the Z136 series may deviate from the requirements of this standard. Each deviation is valid only for applications within the scope of the standard in which it appears. *Guidance in specialized standards (e.g. Z136.3, Z136.4 etc) that appears to conflict with the requirements of this standard shall have precedence within the scope of that standard. The laser safety officer (LSO) shall determine which, if any, of the specialized Z136 laser safety standards are applicable.*

7.9 ANSI control measures

ANSI control measures fall into two broad groups, engineering and administrative. Many engineering controls relate very closely to engineering controls required by laser manufacturers for their laser products. In addition, some controls trace back to early laser systems and are not always easily compatible to present and future laser technology. This text has a separate chapter on ANSI Engineering Control Measures. Control measures are broken down by laser class. As a rule, the overwhelming number of controls apply to Class 3B and Class 4 lasers or laser systems.

The Z136 laser series places a great deal of responsibility onto the laser safety officer, but also gives them a great deal of latitude, from the right to substitute alternate controls to deciding which controls really need to be applied.

7.10 Concluding thoughts

Are there rules that affect lasers? Yes, there are rules for manufacturers and rules for users. If I am a manufacturer of lasers, someone in my organization needs to be the compliance officer and make designers of equipment aware of the requirements and to handle required reporting. As for users, the safety department of the organization one is working for needs to know not only laser use guidance, but what local requirements might be and share that knowledge with users. While for users, standards give guidance but it is following established good practices and the expectation of safety that will really make one safe.

Part II

Training related

IOP Publishing

Laser Safety
Practical knowledge and solutions
Ken Barat

Chapter 8

Safety culture and laser program management

Karen Kelley

8.1 Introduction

It is generally recognized that a positive safety culture is an integral part of a successful safety program. The primary goal of any safety program is to protect the organization's workers. Injuries to workers can have a significant and lasting impact to the workers, the workers' families, and the organization. In addition to the human impact, injuries can lead to financial costs, disruption of operations, damage to workplace morale, lower productivity, and higher turnover. While having a strong culture of safety does not preclude an organization from having accidents, its emphasis on a proactive approach to identifying and mitigating safety concerns can reduce the risk of a serious accident, and when accidents do occur, they are used as an opportunity to learn and improve rather than blame. This first part of this chapter will focus on the relationship between safety culture and the elements of a laser safety program, and the role the laser safety officer plays in fostering a safety culture. The second part of this chapter will focus on safety culture and the incident investigation process.

8.2 Defining safety culture

Safety culture has long been a focus in industries where the consequences of failure are severe, such as the chemical industry, the nuclear industry, aviation, and healthcare. The term safety culture is attributed to the International Atomic Energy Agency's (IAEA) International Nuclear Safety Advisory Group (INSAG), the organization tasked with investigating the explosion at Chernobyl. In their report, INSAG identified 'poor safety culture' as a contributing cause of the accident. The poor safety culture referred to the poor conditions and decision processes in place at Chernobyl at the time of the accident.

While there is not a single definition of safety culture, commonly referenced definitions come from the nuclear industry. The US Nuclear Regulatory Commission's (USNRC) definition of safety culture and the organizational traits of a positive safety culture that they identified can be applied to any industry. The USNRC defines safety culture as 'the core values and behaviors resulting from a collective commitment by leaders and individuals to emphasize safety over competing goals to ensure protection of people and the environment.' The nine traits of a positive safety culture identified by the USNRC are:

1. **Leadership safety values and actions:** Leaders demonstrate a commitment to safety in their decisions and behaviors.
2. **Problem identification and resolution:** Issues potentially impacting safety are promptly identified, fully evaluated, and promptly addressed and corrected commensurate with their significance.
3. **Personal accountability:** All individuals take personal responsibility for safety.
4. **Work processes:** The process of planning and controlling work activities is implemented so that safety is maintained.
5. **Continuous learning:** Opportunities to learn about ways to ensure safety are sought out and implemented.
6. **Environment for raising concerns:** A safety conscious work environment is maintained where personnel feel free to raise safety concerns without fear of retaliation, intimidation, harassment, or discrimination.
7. **Effective safety communication:** Communications maintain a focus on safety.
8. **Respectful work environment:** Trust and respect permeate the organization.
9. **Questioning attitude:** Individuals avoid complacency and continuously challenge existing conditions and activities in order to identify discrepancies that might result in error or inappropriate action.

The process of developing and strengthening the safety culture within an organization can be complex. The laser safety officer may not be in a position to lead broad organizational change, but he or she can help to build a strong safety culture within the laser safety program they manage.

8.3 Safety program maturity

Discussions about safety culture often involve discussions about the maturity of the safety program. There are several safety culture maturity models, also referred to as safety culture ladders. One of the most common models developed by Professor Patrick Hudson (Delft University of Technology) identifies five levels of safety culture maturity from pathological to generative. The table below summarizes the characteristics of each level in terms of safety program elements.

Level	Characteristics
Pathological	*'Who cares about safety as long as we don't get caught.'* Little to no safety program, risk assessment, training, or monitoring. No employee participation. Legal non-compliance. Culture of blame when accidents occur.

Reactive	*'Safety is important after an accident.'*
	Minimal safety program, reactive risk assessment, minimum training, limited incident investigation. Minimal, if any, employee participation. Minimum legal compliance.
Calculative	*'We have systems in place to manage our hazards.'*
	Established safety program and rules based on compliance. Some employee participation. Legal compliance. Causal incident investigation.
Proactive	*'We try to anticipate safety hazards before they arise.'*
	Safety program evolves and improves. Seek to identify problems before they occur and engineer out hazards. Beyond legal compliance. Employee involvement at all levels. Incidents investigated for lessons learned.
Generative	*'Safety is built into the way we work and think.'*
	Risk assessment process fully integrated into all systems. Full employee engagement at all levels. Incidents investigated for root cause and lessons learned.

It is helpful to identify the maturity level of your organization's safety program and where your laser safety program fits into this.

8.4 Relationships as a key element in a laser safety program

The Institution of Occupational Safety and Health (IOSH) identifies three key elements of a positive safety culture in their guide *Promoting a Positive Safety Culture*:
- Working practices and rules for effectively controlling hazards.
- A positive attitude towards risk management and compliance with the control processes.
- The capacity to learn from accidents, near misses and safety performance indicators and bring about continual improvement.

Successfully achieving these elements requires a relationship between the laser safety officer and the laser users that is built on trust, credibility, honesty, and communication. These attributes must flow both ways. The laser users must see the laser safety officer as knowledgeable and trustworthy and a resource and advocate to the laser user. The laser safety officer must see the laser users as subject matter experts in their operations and equipment and an integral part of the development and ongoing management of the laser safety program. It is also important for the laser safety officer to understand and respect the goals and priorities of the laser users and take a genuine interest in what they do and want to accomplish.

8.5 Risk assessments and standard operating procedures

Risk assessments and standard operating procedures are required by laser regulations and standards, but to be effective, they must be seen as more than a compliance checkbox. The risk assessments and resulting controls and procedures are critical elements of a laser safety program and reducing the likelihood of an accident. Many accidents occur when risks have not been fully identified and controlled or when procedures are not followed. To minimize the chance of this happening, the laser safety officer must engage the laser users in the process. Risk assessments and operating procedures are often reevaluated after an accident.

Craig Marriot's book *Challenging the Safety Quo* introduces the concept of investigating when nothing went wrong, learning from normal operations and anticipating and preventing problems before they lead to an accident. Recognize that people are adaptive and innovative and may make minor tweaks to an operation to increase efficiency, effectiveness, and work flow. Some of these tweaks may introduce additional risks. By identifying them during normal operations, the laser safety officer can work with the laser users to develop a better solution. When the laser safety officer has established a relationship of trust with the laser users, it becomes much easier to identify these tweaks. An effective way to identify gaps between the established procedures and practice is to ask the laser users, and utilize their expertize and experience to remove the obstacles to working safely and improve the procedures. Examples of questions include:

- What does your system do?
- What do you see as the biggest hazard of this system?
- What do you see as the most effective way to prevent exposure to that hazard?
- Are there any elements of the standard operating procedure that do not make sense to you or are difficult to implement?
- What can we do to make the system better?

8.6 Training for users and laser safety personnel

Training is another key component of a laser safety program. Organizations with strong safety cultures put an emphasis on providing the knowledge and skills to enable workers to perform their job safely, and they ensure the workers are part of the process rather than just passive receivers.

The laser safety officer may be tasked with developing and delivering training. In order to be effective, the laser safety officer must have strong knowledge of the lasers present in the facility, how the lasers are used and by whom, the applicable standards and regulations, the risk assessment process, and laser controls. Ideally, the laser safety officer has experience working with lasers, but if this is not the case, they will need to gain an understanding of lasers and laser systems. This will improve credibility and the ability to provide effective training.

Laser safety training programs typically utilize several different training methods, including classroom training, online training, hands-on training in a teaching-lab environment, and on-the-job training. Keep in mind that individuals have different learning styles, auditory, visual, and kinesthetic, and combining different training methods will increase the effectiveness of the program as a whole. Hands-on and on-the-job training are especially important components for new laser users to allow them to demonstrate their skills and proficiency before working without direct supervision.

Training programs should not be static and should be updated based on ongoing needs assessments and feedback from laser users.

8.7 Change management

Even seemingly minor changes to a laser system or the environment can lead to unforeseen hazards. When a strong relationship exists between the laser safety officer and the laser users, changes can be communicated before they are made, new hazards, risks, or requirements can be identified, and controls and procedures can be updated. The laser safety officer should make it a habit to ask laser users if anything has changed or if any changes are planned.

8.8 Continuous improvement

Organizations with a strong safety culture support an ongoing process of continuous improvement that involves identifying and reporting unsafe acts, conditions, and near misses, and implementing change proactively to prevent accidents from occurring.

The laser safety officer's relationship with the laser users is critical in establishing this process of continuous improvement. Laser users need to feel comfortable reporting unsafe acts, conditions, and near misses to the laser safety officer. The laser safety officer's response must be non-punitive, and he or she must engage with the laser users to identify appropriate improvements.

Some organizations have formal processes for reporting and investigating near misses, but even with a process in place and a culture of non-punitive responses, near misses are still often underreported. One reason for this is a lack of understanding of what a near miss is and how to identify it. A near miss is an event that had the potential to cause injury or damage if the circumstances were slightly different. An example of a near miss would be a stray beam generated during an alignment procedure entering another area where personnel were working. The key benefit of investigating near misses is to identify and implement improvements before an accident occurs, and to use this as an opportunity for a lesson learned across the broader organization.

8.9 Incident investigation

As mentioned earlier in this chapter, having a strong culture of safety does not preclude an organization from having an accident, but its emphasis on a proactive approach to identifying and mitigating safety concerns can reduce the risk of a serious accident, and when accidents do occur, they are used as an opportunity to learn and improve rather than blame.

Organizations with strong safety cultures will have an incident investigation and analysis process that is driven by a deep understanding of how accidents happen. Incidents are rarely the result of a single malfunction or action, but rather the result of multiple system or organizational failures. James Reason's 'Swiss cheese model' illustrates this concept by representing the multiple layers of defense as slices of Swiss cheese and the holes within each slice represent opportunities where failure can occur. When the holes align, an incident results.

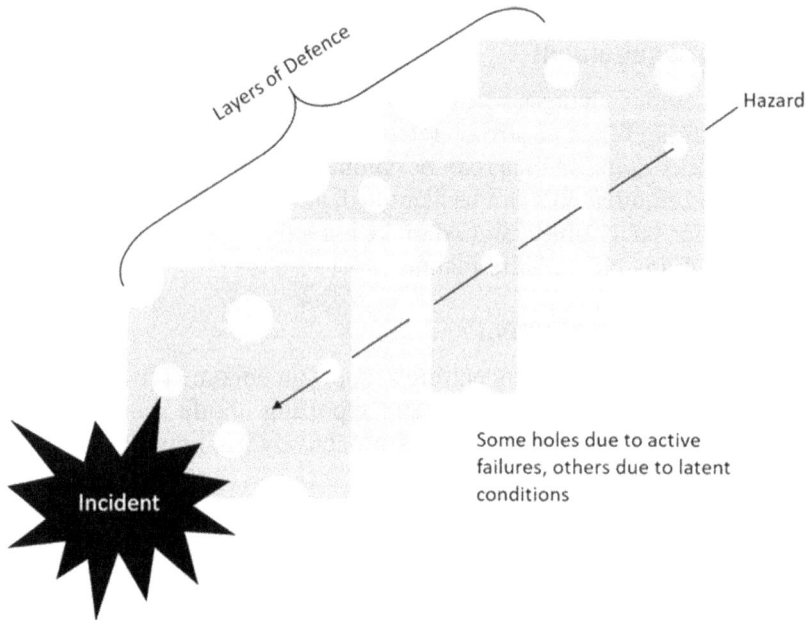

Layers of Defence

Hazard

Incident

Some holes due to active failures, others due to latent conditions

When a laser accident occurs, the laser safety officer should lead the investigation to identify all of the gaps and failures that contributed to the accident and focus safety improvements at each of these levels in order to have a greater impact in preventing future accidents.

8.9.1 Root cause analysis

There are a number of incident investigation methodologies, and a specific methodology may be chosen based on the severity of the incident. The laser safety officer should familiarize themself with the different methodologies and choose the one that best fits the situation. Regardless of which method is used, all incidents should be investigated to identify all of the causes of the incident:

- **Direct causes** are the aspects of the incident that immediately contributed to the injury or damage. An example of a direct cause of a laser eye injury is a specular reflection from a sensor card striking the eye during an alignment procedure.
- **Contributing causes** are the symptoms, the unsafe acts or conditions that contributed to the incident. An example of a contributing cause of a laser eye injury is the failure to wear laser protective eyewear during alignment.
- **Root causes** are the underlying system, management, or organization related failures that led to the unsafe acts or conditions. An example of a root cause of

a laser eye injury is the failure of management to consistently enforce the laser protective eyewear requirement, leading the laser user to believe that wearing eyewear is not really that important. A step further would look at the failure of management to engage the laser user in finding a better way to protect from the hazard, possibly eliminating the dependence on laser eyewear.

Many incident investigations stop at the contributing cause, the unsafe behavior or condition, without identifying the underlying causes. It is the identification and correction of the root causes that led to the decision to behave in an unsafe manner or ignore an unsafe condition that can lead to the most impactful and long lasting changes.

As previously mentioned, the incident investigation method may be chosen based on the severity of the incident. The method may also be chosen based on the type of incident investigation that will be conducted, such as a one-on-one discussion or a more formal team-based investigation. A simple incident investigation method is the '5 Whys'. The 5 Whys is a technique used to determine the root cause by repeating the question 'Why?' at least five times. This method may be more suited for minor incidents or one-on-one discussions.

More serious or complex incidents may require a more sophisticated investigation technique, such as a causal tree. As described in the Center for Chemical Process Safety (CCPS) *Guidelines for Investigating Chemical Process Incidents*, the causal tree method starts at the end of the event and works up the tree one level at a time, asking the following questions at each level:

1. What was the cause of this result?
2. What was directly *necessary* to cause the end result?
3. Are these factors (identified from question 2 above) *sufficient* to have caused the result?

There are a number of resources available for laser safety officers to familiarize themselves with incident investigation methodologies.

8.9.2 Team-based investigations

Incident investigations can benefit from a team-based approach, particularly when investigating more serious or complex incidents or incident trends. Team-based investigations are also valuable when the laser safety officer does not have direct laser experience.

The CCPS *Guidelines for Investigating Chemical Process Incidents* lists several advantages to using a team approach for investigating incidents, including:

- Multiple technical perspectives assist in analyzing the findings. A formal analysis process must be used to reach conclusions, and individuals with diverse skills and perspectives best support this approach.
- Diverse personal viewpoints enhance objectivity. A team is less likely to be subjective or biased in its conclusions, and the conclusions are more likely to be accepted.

- Internal peer reviews can enhance quality. Team members with relevant knowledge of the analysis process are better prepared to review each other's work and provide constructive critique.
- Additional resources are available. A formal investigation can involve a great deal of work that may exceed the capabilities of one person.

8.10 Conclusion

The safety culture of an organization has a significant impact on a laser safety program. While the laser safety officer may not be in a position to bring about broad changes in the organization's culture, he or she can help to develop or improve the safety culture by establishing relationships of trust and respect with laser users, and engaging the laser users in the laser safety program management and continuous improvement processes.

IOP Publishing

Laser Safety
Practical knowledge and solutions
Ken Barat

Chapter 9

Performance-based learning in laser safety training

Ken Barat

9.1 Introduction

Laser safety training for users of Class 3B and Class 4 lasers has been a cornerstone of laser safety from the start. Institutions and professional laser safety training firms provide well-crafted laser safety courses, but the training method has not significantly changed in more than 40 years, it is either lecture or web based. There have been modifications such as embedded videos and animated PowerPoint effects, but, with limited exceptions, all have a passive lecture format. Problem-based learning was originally developed for use in medical education and was quickly adopted by schools of business. It has only recently become more prevalent in technology and engineering education because it teaches students the process of solving open-ended problems at the same time as they are learning new material. One of the most famous applications has been engineering design classes at MIT. PBL can be applied to Laser Safety Officer and User training and can have significant rewards. PBL exercises can be brought into present training modes. How much does one take away from the overdose of facts in a standard training course, whether it is a few hours or several days? PBL allows you to try out solutions to beam control challenges in a stress-free environment.

In the field of education two tools receiving a great deal of attention are performance-based learning (PBL) and problem-based learning (PBL). Both approaches have been shown in several studies to increase concept retention and the ability to transfer knowledge and skill to new situations long after training is complete. This discussion centers on performance-based learning for the Laser User and Laser Safety Officer. An article by Sierra Training Associates, stated that the researcher found that students immediately after the PBL process, did significantly poorer than traditionally instructed groups on immediate multiple choice tests. Six months later, though, these same PBL students demonstrated a recall of concepts up to five times higher than the traditionally instructed groups. The above findings are clear reasons why PBL should be used in basic Laser Safety Officer (LSO) and laser safety training as often as possible. The role of the LSO is to use professional judgment. This is one of the advantages and strengths of the standards. An example is the inclusion of 'Substitution of Alternate Controls', found in Z136.1, Z136.8 and Z136.9.

9.2 Alignment demonstration

After an accident some US Department of Energy facilities have either established or evaluated a hands-on laser alignment class. It needs to be clear to the reader, these classes are not prolonged courses, their duration is only a few hours, they cannot replace semester long optics handling training. Rather they are a set of exercises for a student to do, such as placing a visible beam through a series of irises or periscopes. Follow up examples of such a class will be found at the end of this chapter.

More examples

A general technique or flow for PBL is the following:
- Individuals (or groups, groups work best) are given a problem to solve.
- The problem should relate to anticipated work and be based on real world situations.
- The problem should not offer easy or clear solutions.
- Learners must identify possible needs, locate potential resources and use them.
- Problems should drive the need to learn more as a process of resolving it.
- Learning is initiated by the problem.
- Learning is active and integrated.

An effective problem should have these elements:
- Require thought and decision making by the students.
- Multi-stage, with information unfolding from stage to stage.
- Designed to encourage group problem solving.
- Aligned with course learning objectives.

PBL is ideally suited to adult learners like those found in a typical laser safety course. Compared to younger students, adults are more self-directed and learn best when new knowledge builds on their prior knowledge and life experiences. Despite the suitability of PBL to this audience, the first PBL exercise should include a 'walk through' of structured problem solving so that time is used most efficiently in subsequent problems. In its most basic form, structured problem solving requires:
- Clearly defining the problem and identifying the characteristics of a successful solution.
- Determining what knowledge group members have and what they need to learn to solve the problem.
- Dividing up the task of acquiring the needed knowledge.

- Group brainstorming to determine the best solution.
- Testing the solution against the characteristics listed in step 1 to ensure a successful solution.

In this student-centered method of instruction, the instructor becomes a facilitator, providing targeted instruction on an as-needed basis. Students ask questions because of a 'need to know' in order to solve the problem. They are more likely to pay close attention to the answer (and thus retain the information) because the answer fits into a solution they are crafting with their team.

Special Note: Performance-based learning is the same as competency-based learning where a student is assessed according to what he/she can do (i.e. align optics, put a beam through three irises, measure voltage with an oscilloscope, etc). The focus is on what a student can do and demonstrate. In general, it's not the same as problem-based learning, where new material is mastered in the context of solving an authentic problem.

PBL is an attempt to get the student involved in the learning process and increase retention of the learning material. It is the contention of the authors that laser safety lends itself well to PBL. PBL does take more time and depending on the approach requires more preparation by the students. In the typical 3–4.5 day laser safety course, it is quite common for the instructor to barely have sufficient time to get all the material across, in particular if the class is involved and asks questions. However, simply 'covering the material' does not guarantee that students have learned the subject matter in a deep way that will allow them to apply it to new situations. With some pruning of material, the laser safety instructor can add PBL elements.

9.3 Examples that apply PBL to laser safety training

So where can PBL be added to an LSO training course? Consider the following as a PBL exercise:

9.3.1 Eyewear selection

Give students specifications of lasers used in a work location. Let them calculate the optical density and choose the correct filter from a list of filters. If time for calculations is not sufficient or not part of a course, tell them the wavelength and optical density needed and allow students to pick from the filter selection.

Then add information such as lighting conditions, which beams are most accessible, is alignment eyewear available, is there a one filter that fits all or how many pairs of eyewear are required?

9.3.2 Service of Class 1 product

Give the students the layout of a room, make sure it has windows to the hallway. Describe it as a biotechnology lab, one that contains Class 1 products, such as a cell sorter, gene sequencer, Raman unit etc. Now, ask them when the service person opens up the unit and Class 4 laser radiation exposure is possible, how they would establish a safe work area. The same can be done in a more industrial setting with, say, a micro-welder or additive manufacturing (3D printer) system.

9.3.3 Control area challenge

Set up a work scenario, laser, wavelength, power, basic layout. Ask the students to develop standard operating procedures for safety in normal operation. How would controls change during alignment and service? Use various scenarios: on a factory floor, a research lab, an open user facility, depending on the needs and interests of the students in the class.

9.3.4 Accident review

Give the setting of a laser accident, ask the group what they found wrong and how to correct it. One interesting way to do this is for one or two instructors to play the role of the victim or a group of staff and for them to be interviewed by the student group to determine what happened and how it could have been prevented.

9.3.5 Lab design

Give a group type of lasers and basic layout and ask them to design a lab, showing safety and basic room requirements. Then ask how the design would change if it were a factory or large open space user facility.

The role of LSO and/or laser user requires professional judgment. Even in the world full of turn-key and systems equipment integration, laser safety requires knowledge in order for LSOs/users to feel confident about their decisions. This is even truer in the research setting where access to beams is the rule rather than the exception. Poor decisions by LSOs or users can put people at risk or cause life changing injuries. So any way we can have students leave a training class with real confidence that they can address problems they will find in their work environment, rather than just a head full of facts, must be our desired goal. It is time to apply performance-based learning to laser safety training.

9.4 Value of these PBL exercises

The real value of these exercises is that in a non-stress atmosphere the LSO or student can review options, see what other considerations might exist or affect their decision. If nothing more it will show that nothing is as simple as it seems.

9.5 Alignment class material

The following is material that has been applied to a hands-on laser alignment class of limited duration.

Here is sample material from such a class:

- Hands-on laser safety and alignment laboratory.
  - Pre-class activity by trainees.
  - Safe alignment procedures demonstrated by lab supervisor.
  - Hands-on alignment of visible and invisible Class 1 lasers.

- Fundamentals.
- Remove reflective accessories.
- Beam plane below eye level.
- Keep eye out of beam path and plane.
- Perform alignment at low power.
- Perform mini-sweep at each optical element.
- Block beam before adding optical elements.
- Use barriers to confine beam to optical table.
- Perform final sweep to verify beam control.

Trainee demonstrations:
- Locate and block all unwanted reflections at each optical element.
- Set up barriers to keep beams confined to the area of table designated for their alignment exercise (figures 9.1 and 9.2).
- Discuss alternate techniques with instructor.
- Perform complete sweep to ensure beams are confined to the optical table.

Desired training outcomes:
- Remove reflective accessories.
- Beam plane below eye level.
- Keep eye out of beam path and plane.
- Perform alignment at low power.
- Perform mini-sweep at each optical element.
- Block beam before adding optical elements.
- Use barriers to confine beam to optical table.
- Perform final sweep to verify beam control.

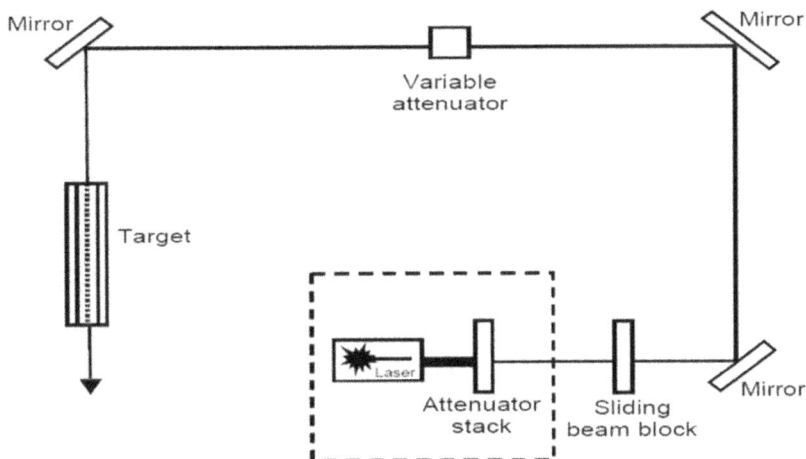

Figure 9.1. Visible beam alignment exercise.

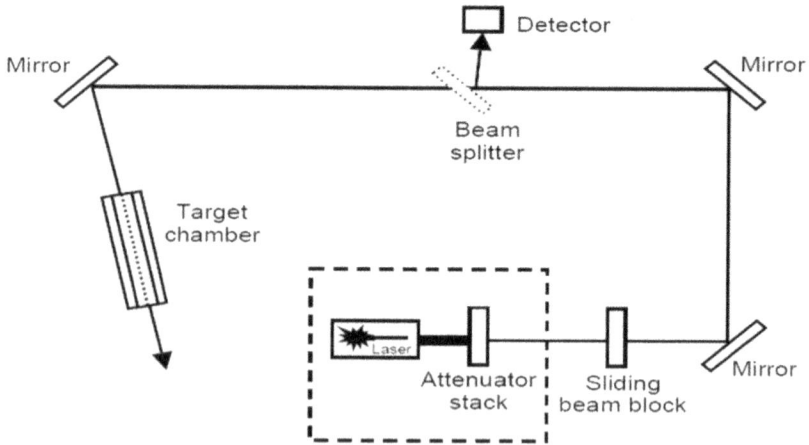

Figure 9.2. Invisible beam alignment exercise.

The results

Such classes can give a supervisor an idea of the optic handling skill level of a new employee. Armed with that knowledge the supervisor can see where mentoring is needed and, if necessary, what restrictions need to be placed on the individual's activities until the skill level increases.

9.6 Conclusion

Performance-based training is a way to augment traditional PowerPoint or web training. It also helps the student retain material knowledge and aids in problem solving. It is well worth the effort to find ways to include it in your training.

Chapter 10

Training, breaking through to users

Ken Barat

10.1 Training

All users of Class 4 laser systems require some laser safety training. The training can be broad or specific to the hazards they will encounter with their system. The overwhelming number of laser standards and regulations call for what I would term institutional laser safety training. Which is really an awareness of laser hazards and safety basics.

Of course, almost no-one is placed in a Class 4 environment without some additional on the job training. It is that site specific training that is really going to keep one safe. While few do, it is prudent to document that training. Some folks also refer to this as micro training or mentoring.

Lastly is the idea of refresher training as a booster shot to keep users aware of laser safety practices.

Each of the three training types (institutional, on the job and refresher) has its own delivery style and goal. All should be reviewed periodically to ensure they are effective and current. In addition, each of the training types mentioned have built-in traps that can detract from the purpose of the training. So rather than reviewing ways to do the training, let's talk about what to avoid.

10.2 Institutional—required by regulations and standards

This can be too broad, especially if one uses any of a number of commercial 'canned' courses. Not only can these spend time on topics of no interest to those taking the training but can also raise unnecessary concerns and fears and gloss over relevant items of the users. Therefore, it is important to evaluate what information is 'nice to hear about' and what information is 'necessary' for the audience.

10.3 On the job training—site/equipment specific

The most common trap here is to give this task to the most ill prepared person on the existing staff. The trainer needs to fully understand their responsibility. The training needs to follow a set course and have a set number of documented goals. In addition, there should not be a time limit allowed for such training. How long it takes will vary with the trainee's experience, language skills, comprehension and willingness to ask questions, as opposed to just shaking their head in agreement.

10.4 Refresher training

The questions of how often to do this and how extensive it needs to be are not easy ones to answer. Much depends on the experience of the trainee, how much, and how often, they have contact with the systems. The usual errors are to give the refresher training too often, making it too long or too short. Repeating the same training each time is also a way to defeat the training's purpose. Variation and even group discussion of the work area controls can be more effective than a set presentation.

10.5 Service staff challenges

No-one can undermine a safety program more than an outside service person. They are looked upon as the system expert. If they do not wear eyewear, post signs or follow good safety practice, when observed by your staff they will ask 'Why should we do these things, if the system expert feels they are unnecessary?' Most outside service people do not carry with them signs or temporary barriers, they rely on you to provide them. This is a real problem with Class 1 products, that when open become Class 4. The simple reason is, when running as Class 1, no-one thinks or should be thinking about laser safety. However, all that changes when doors and panels are open and the lasers are running.

There is nothing wrong with expecting outside service staff to follow your safety practices, in fact they should not be allowed to work if they do not.

10.6 Mobile app training, an effective training delivery approach

The use of mobile apps is not completely new. More and more safety related items can be found on people's smart devices. Emergency response is one that is gaining popularity. So, providing short training modules on people's mobile devices is in line with that. This technique can easily be a way to remind people of safety or a mentoring technique. It can be text or video. An example might be a video on the proper technique to clean a lens. Just as how Twitter is used for fast communication, lessons learned, or other training items can be distributed this way.

10.7 Microlearning?

This is a buzz word people will be hearing more and more. The goal is to deliver training in short bursts or pulses. It fits well with the mobile app topic above. But it can also be a live, person-to-person event, hey mentoring! Microlearning can also be a follow-up to the traditional, full-day, safety orientation in which there is such an overload of information; often the only goal achieved is presenting the information, not comprehension.

10.8 Game learning

This is a combination of the mobile app and microlearning, where the lesson is hidden within a puzzle of a competitive quiz. Anything that will engage your staff and help them remember safety practices is worth a try.

10.9 Just-in-time learning, safe plan of action

Just as 'just-in-time inventory' has become an accepted practice for many firms, we need to remember that when a new activity or system arrives, new training may be needed. This is not the same as training from a manufacturer on how a piece of equipment works, but covers what new safety concerns need to be addressed. This is why the safe plan of action, tailgate style meeting with staff on the new system has great value. A discussion of how things will work, do we have all the required tools for safe work, is more training needed and where do the hazards exist (such as potential for reflections), etc?

Note: See chapter on performance-based learning.

10.10 On the job training—OJT or is it just mentoring?

On the job training (OJT) or a more upscale name for mentoring is of vital importance for a new employee or grad student. Not only does this activity ensure they know how to perform the required tasks, but it helps them understand your safety culture, work expectations and allows you to see if their level of experience is as advertised. Institutional laser safety training, or any institutional safety training, at best really just raises staff awareness of hazards. It is site specific OJT that will make and keep them safe.

Few employers or researchers will turn over 100 000s of dollars' worth of equipment to someone and say 'go for it, you know what to do'. Unfortunately, many times the importance of the mentor or OJT training is overlooked. The task is turned over to the individual whose time is deemed least valuable, which is not a proud title to own. In addition, for many, it is a race to see how soon this critical job can be performed so they can get back to their own work.

The following sections are some topics related to OJT.

10.10.1 How long should it take?

The quick answer is as long as necessary, based on the individual's previous skill set and learning curve. Some institutions set a pre-determined mentoring period. This has pros and cons, if a reasonable time period is selected. In addition the answer to the question is dependent on the task the trainee is being trained for. If it is training to operate a set piece of equipment, then the mentoring period can be rather short. If it is training a student who will have a great deal of independence once OJT is completed, the length of time can be extensive. It may be broken down into steps or approved tasks.

If a prolong period of say 30 days is set, all have an expectation of completion as well as the pace that may be set. This allows for greater time to be spent on explanation of activities as well as observational periods. Observation is both how the work is done and how the individual does the work. The time people perceive as being lost to OJT will be reward by superior performance and less costly errors.

If too short a period is set, the pressure to get it done could lead to compressing the training and reducing it to just what the mentor considers 'the skill of the craft', or just the essential knowledge the mentor believes the person must have.

So, in conclusion, the answer to the question of 'how long should it take?' is: a time period should be set, based on previous experience of providing OJT or in particular in the case of an academic setting until the mentor is completely satisfied the student can perform the work unsupervised and is willing to sign off on that fact.

10.10.2 How should OJT be performed?

OJT has several steps, simply put it involves: explanation of the task, demonstration of the work/task/operation, observation of the individual doing the task, answering and soliciting questions from the trainee and making the trainee feel comfortable asking questions. For those of you who learned how to ride a two-wheel bike it might have just been: get on, keep your balance and go. Even in such a cryptic OJT someone might have been running alongside you for a while. In today's techno-logical world we are looking for a little more. The explanation of the science, definably an understanding of the hazard and safety steps, as well as the how to do it (what buttons to push). When you get to the step where the mentor is ready to sign off to allow unsupervised work, the trainee also needs to sign indicating they feel they are ready.

10.10.3 What needs to be covered?

OJT should not follow a cookbook approach, add a cup of this and tablespoon of that and stir. If I am giving OJT to a new employee or student in my work area, I want to cover the things that make life better and safer. Typically, new employee orientation is a *blitzkrieg* of information, one generally acknowledges receiving but little is retained. OJT in the work area needs to cover a few site specific points, usually overlooked but extremely important to safety. What is the evacuation route and assembly point, where is the fire extinguisher, eyewash shower? Where can emergency response information be found and, of course, where is the best place for lunch.

10.10.4 Who should the mentor be?

Many times, there is no choice on the selection of the mentor or trainer. There is only one other person in the research group or only a few people who know how to operate the equipment. The most important thing is that, whoever the individual is, they need to know how important their role is. The passing on of poor safety habits, for example, is a major issue in establishing a workable safety culture. Not wearing eyewear or wearing it on ones forehead during alignment is not a positive message. Controls that have been put into operating procedures just to satisfy the Safety Department but which are not followed is a danger and legal liability for all parties. The faculty member who turns over his lab operation to a Post Doc or student without upholding a level of safety compliance is only putting their heads in the sand waiting for a problem to bite them. The same is true for the faculty member who is so busy teaching and writing grants they can only hope the student was a good choice and hope for the best. So who should the mentor be? The answer is someone who will take the task and responsibility seriously. Remembering the students' performance reflects on them also. They can be the root of a very productive and successful series of experts or the source of many subpar performers.

10.10.5 Do as I say not what I do

How often have we all heard this or said it ourselves. What it means is 'I know this is the proper or safe way to do something, but this is faster or almost as safe'. Just like stepping on the top level of a ladder, right over the sign that says do not step here. This all becomes a matter of safety culture. If a procedure is unclear on why it should be followed either change it or see that all understand its reasoning and acknowledge it.

10.10.6 Does OJT need to be documented?

People take things seriously when they have to put their signature on the line. OJT needs to be documented. Both the mentor and trainee should sign they are in agreement that the OJT has been successfully completed (a sample form is shown). Even if the training is done in a piecemeal fashion, as each task is learned a sign off should take place. In addition, documented OJT helps protect the mentor if training topics are outlined showing a history of what was instructed. So no-one can say 'I was never told *don't* push that button'.

10.10.7 Is follow-up needed?

Once a trainee has been working unsupervised for a period of time, as a point of good practice, a follow-up check by the mentor can be of great value. It is at this time we look for those follow-up questions that demonstrate growth; or sometimes items that show a lack of understanding. Do they have suggestions on how to improve working practice, what steps have they taken to do this, are things running smoothly and they are looking for more improvements? All these questions and observations give the mentor insights into the person they hired and how they will progress in their role. Their progress also demonstrates the mentor's success, for remember, the trainee's performance reflects on you as well as them.

In conclusion: Mentoring, or on the job training, is not only critical to job performance and safety but also employee retention. Studies have shown that employees who feel the employer is invested in their success stay longer. Mentoring is one of the best and easiest ways to show that investment. Answer the question, what is the legacy you wish to impart to your staff? Lastly, those you mentor reflect upon you, what is the reputation you wish to develop?

10.11 More on refresher training

The concept of refresher training in laser safety is not new. Refresher training is brought up in both ANSI Z136.1 Safe Use of Lasers and ANSI Z136.8 the Research Laser Standard and Z136.9, the Industrial Laser Standard. Many institutions perform refresher training, but questions of how often, what to include, and how to deliver remain. The following are some bullet points on the topic.

10.11.1 Why refresher training?

- We all know how easy it is to get caught up in our work.
- During that process, it is also easy for negative habits to creep into our routine.
- A means to reintroduce good work practices is essential to keep incidents from happening.
- That is what I am terming refresher training.
- It is a good work habit re-enforcement program.
- Lessons learned and best practice programs also help.

10.11.2 How do existing standards address the idea of refresher training?

- Z136.1 Shall consider.
- Z136.2 Not addressed.
- Z136.3 Not addressed.
- Z136.8 Shall address need for (meaning could or could not do).
  - Pending version-shall provide.
- Z136.9 Shall provide.

Note: IEC 60825-14, does not address refresher training.

10.12 Effective refresher training what are one's options?

- Repeat initial training.
- Dedicated refresher course.
- Quiz out.
- Group specific training, LSO led.
- Group specific training, self-led.
- Peer refresher.

10.12.1 Frequency is an important factor

- We do not want people to feel, that every time they turn around the training is due again.
- Annual, too often and a burden on all parties.

- Every two years, like radiation safety refresher training, also too often in my opinion.
- Every three years, like Goldilocks, just about right.
- Every 4–5 years, too far between sessions.
- Too short a time frame creates several problems for the institution.

The biggest challenge is keeping the training fresh and relative to users, more later on this, so do not leave:
- Refresher training must be relevant.
- It must be relevant to the population.
- It must be fresh, not the same material every year.
- It should involve some work where the trainee must get involved.

10.12.2 Refresher conclusion

Refresher training has an active role in keeping a laser safety culture alive. It cannot be just a 'box ticking' exercise. A successful and performance-based problem does take work and thought. Your decision should be that you are willing to see that it will work at your institution. Management must allow the LSO the time to perform effective laser safety refresher training.

10.13 Conclusion

No matter how it is achieved, Class 3B and Class 4 laser users are going to receive some laser safety training. The goal is not to put them to sleep, but to find information and ways to deliver it that will make an impression and want and help them work in a safe way. No-one but a few masochists want to be injured. Our goal has to be, to light the fire under that desire and keep all safe.

Chapter 11

Mentoring, do what I say and follow my lead

Ken Barat

11.1 Introduction

Most children learn a great deal from their parents. Sometimes good habits and perspectives, other times well let's say not so good. Mentoring, or on the job training (OJT), concerning work habits and procedures in the laser lab is the same. OJT and mentoring will also be addressed in the chapter on training—its importance and impact on laser safety cannot be over-stressed, so it will be addressed from different points of view in each chapter.

Experience has shown that some activities are never taught and are thought to be inherently known. Many of these fall under the title of 'skills of the craft'. If you are a journeyman electrician, you are expected to have a certain knowledge base. Unfortunately, skill of the craft, just like common sense is many times expected, but not present.

Items like how to use an IR sensor card, how to use an IR viewer and the use and handling of a power meter are examples. While seemingly obvious, each has a right and wrong way to be used and applied. Reflections of plastic IR cards is a good example.

The laser lab does not have a flight simulator, where the effects of bad habits or skills can be shown with no negative effects. As technology advances, the increased use of virtual reality learning systems may fill this gap. I have come across two or three of these systems in actual use and development.

So, back to mentoring. As with most activities there is a right way and wrong way. Before we get to techniques, I want to emphasise the need to document OJT. Something about signing one's name on a form makes one more responsible. Not the same as clicking 'I agree' to terms on a web page. Just think how many times you have clicked on privacy policies but never read them, just to get on to what you want to do.

The OJT form needs to be signed by the trainer and trainees. Indicating that both agree that the training has been sufficient for the trainee to work unsupervised.

doi:10.1088/2053-2563/ab0f25ch11

11.2 Goal of mentoring

The goal is to impart or train a person with the skills to be successful in their field and the techniques that go along with that. It is necessary to have an understanding of how things work and why they are chosen.

The mentor should not assume the trainee knows things until demonstrated. Always remember the trainee will one day be the trainer and we want good practices and techniques passed on, not bad ones (which are the easiest thing to pass on). In addition, as a well-worn saying goes, 'actions speak louder than words'.

11.3 The 10 core laser safety principals

In addition to task specific activities and skills, the following 10 items should also be addressed in any OJT or orientation to working in a laser environment. The trainer should also take this time to go over the location of fire extinguisher, eye wash showers, the evacuation route and assembly points and emergency exits.

1. Selection of proper eyewear.
2. Checks on the condition of eyewear.
3. Alerting others prior to turning on laser and of open beams.
4. Thoroughly checking for stray reflections.
5. Blocking stray reflections.
6. Demonstrating beam detection methods.
7. Understanding controls for different intensity levels.
8. Reading and familiarization with controls per SOP.
9. Familiarity with equipment safety features.
10. Communication with others.

11.4 How to be a good mentor or trainer

If one looks for good practices on mentoring, a great deal of advice is available in articles, textbooks and web pages. Looking over these, here are some of the good skills most frequently listed.

A detailed list of mentoring skills and approach
- Active listening is basic to mentoring.
- Be encouraging.
- Ask questions.
- Expect setbacks, not perfection.
- Provide corrective feedback.
- Ability and willingness to communicate what you know.
- Be prepared for each day.
- Approachability, availability.
- Objectivity and fairness.
- Compassion and genuineness.

11.5 For those that like it short and simple

1. Mentoring the individual through instruction on the work process and safety steps.
2. Demonstrating hands-on skills.
3. Observing the individual perform the activity.
4. Receiving feedback from the trainee.

11.6 A hard lesson for those giving OJT/mentoring

11.6.1 Hold direct answers back

You will expect technical questions and it will be easy to give answers. Hold back. It is well documented that people learn a lot better when they figure things out for themselves. Rather than giving the answer, steer them in the right direction by asking questions or suggesting other ways of approaching the problem.

Of course some questions do not lend themselves to this approach but require a straightforward answer.

11.7 What about mistakes?

People can learn from mistakes and sometimes mistakes provide opportunities for the best learning experiences. So without hurting anyone or breaking valuable equipment, allow that mistakes may happen by the trainee. How the trainer responds to mistakes will have a major effect on how well the training goes from that point on.

11.8 Commonly overlooked topics

- How to hold a sensor card.
- Use of periscope.
- Unofficial beam blocks (paper or other materials).
- Power meter movement/adjustment.
- Use of reflective tools.
- Reaching across the table.

11.9 Safety culture

The term laser safety culture over the last 10 years has received a great deal of attention, prompting questions such as: 'what is your safety culture?', 'how do you establish a safety culture?' and 'how do you keep a safety culture going?'.

In simple terms, a laser safety culture, or any safety culture, is an expectation that safe work practices, such as wearing eyewear or enclosing beams is the norm. Deviations will be brought to one's attention and will be corrected. Of course, for many institutions and work sites, safety compliance/culture is strictly an individual choice. I certainly know places where safety is down to one word 'Duck!', as long as one does not get hit by the beam one must be working safely.

No one should accept such a philosophy as the norm. If I saw a sewer opening on the street, I would need to cover it or put a barrier around it. I am not saying many would not notice the hole and step into it, but betting that someone will is a bet I would take.

This text contains a great deal on mentoring and on the job training. What we need is group peer pressure that safe practices are our expectation. This expectation must hold during normal operation and especially when project pressures arise and time is critical. Which, of course, is the most common time for safety steps to be set aside.

So let's just take a look at how others define safety culture.

According to OSHA, '*safety cultures* consist of shared beliefs, practices, and attitudes that exist at an establishment. *Culture* is the atmosphere created by those beliefs, attitudes, etc, which shape our behavior.'

Here are a couple of tips from OSHA to get you started on building a strong safety culture at your organization:

1. Define safety responsibilities: Do this for each level within your organization. This should include policies, goals and plans for the safety culture.
2. Share your safety vision: Everyone should be in the same boat when establishing goals and objectives for their safety culture.
3. Enforce accountability: Create a process that holds everyone accountable for being visibly involved, especially managers and supervisors. They are the leaders for a positive change.
4. Provide multiple options: Provide different options for employees to bring their concerns or issues full-face. There should be a chain of command to make sure supervisors are held accountable for being responsive.
5. Report, report, report: Educate employees on the importance of reporting injuries, first aid and near misses. Prepare for an increase in incidents if currently there is under-reporting. It will level off eventually.
6. Rebuild the investigation system: Evaluating the incident investigation system is critical to make sure investigations are conducted in an effective manner. This should help get to the root cause of accidents and incidents.
7. Build trust: When things start to change in the workplace, it is important to keep the water calm. Building trust will help everyone work together to see improvements.
8. Celebrate success: Make your efforts public to keep everyone motivated and updated throughout the process.

The United Kingdom Health and Safety Executive makes it clear that from their point of view you cannot have a viable safety culture unless management buys in. Something any safety person will tell you. Their signs of a strong positive management culture are:

1. Managers regularly visit the workplace and discuss safety matters with the workforce.
2. The company gives regular, clear information on safety matters.
3. We can raise a safety concern, knowing the company take it seriously and they will tell us what they are doing about it.

4. Safety is always the company's top priority, we can stop a job if we don't feel safe.
5. The company investigates all accidents and near misses, does something about it and gives feedback.
6. The company keeps up to date with new ideas on safety.
7. We can get safety equipment and training if needed—the budget for this seems about right.
8. Everyone is included in decisions affecting safety and are regularly asked for input.
9. It's rare for anyone here to take shortcuts or unnecessary risks.
10. We can be open and honest about safety: the company doesn't simply find someone to blame.
11. Morale is generally high.

All safety professionals and upper management say they want a positive safety culture. As mentioned in the OJT chapter and throughout this book, an expectation of good practice safety needs to be an individual decision. Management can set expectations, but the individual makes the difference. People have been successful who are about as unsafe as possible, safe practice is not a guarantee of success. But when training those coming after us, it is a moral duty to teach how to work safely and show how safe practices are not a barrier to successful results.

Chapter 12

Can everyone understand your work? Considering visual disabilities when designing graphics and presentations

Lisa Manglass

12.1 Introduction

12.1.1 Disability and ethics

A modern perspective of disability considers disability to be a concept that results from barriers that prevent some people from full participation in society as opposed to seeing disability as a feature that describes a person. In 2006, the United Nations general assembly adopted the Convention on the Rights of Persons with Disabilities, a framework that includes recommendations based on the principles that people with disabilities have the right to both 'non-discrimination' and 'full and effective participation and inclusion in society' [1]. Scientists and safety professionals who strive to maintain a code of ethics should consider disability at many phases of their work, but especially in how they communicate science and safety. Taking time to consider visual disabilities when you present your work promotes inclusion with minimal effort.

12.1.2 Types of visual disabilities

The human eye is a complex structure of fluid, blood vessels, and nerves that send information from a receptor to the brain. Simply put: there's a lot that can go wrong. When most people think about a visual disability, they think of total blindness, but there is a large range of visual impairments. Impairments such as severe myopia that cannot be fully corrected with glasses or contact lenses, cataracts, or reduction of visual field may impact the way that a person interacts with the world, however, it does not mean that they don't benefit from their vision in some capacity. People with severe visual impairments/low-vision or total blindness account for an estimated 3.3% of the population in the United States [2] and 4.23% world-wide [3].

Additionally, about 8% of male and 0.5% of female people have some type of color vision deficiency.

12.1.3 Why do we care?

Visualization is one of the most powerful tools that researchers and safety professionals use to effectively communicate their work. Visualizations can allow us to describe a large quantity of data at once. They can also effectively and quickly communicate a message. However, when visualizations are created without regard for any visual disabilities, we may be ineffectively communicating with close to 10% of our audience. While those with visual impairments are a minority of the population, 10% will still represent a large number of the people attending your talk, reading your paper, or working in laboratories that you supervise. There is no single, easy solution to making your message or presentation ideal for everyone but making a few simple considerations in how you present your work or message has the potential improve your ability to communicate much more effectively with your audience.

12.2 Color deficient vision

12.2.1 Color vision

A human eye contains photoreceptors that are responsible for absorbing different wavelengths of light called cones. There are three different kinds of cones in a typical human eye, where each cone is roughly responsible for long, medium, and short wavelengths of the visible light spectrum, conventionally referred to as red, green, and blue, respectively. The typical eye contains about 45% each of red and green cones and 10% blue cones. Despite the conventional color names for these cones, each is responsible for a range of wavelengths, and there is a large amount of overlap in the wavelength ranges of the green and red cones. The information gained from these receptors is interpreted by the brain by intensity and by the ratios of output from cones relative to each other [4].

12.2.2 Types of deficient color vision

Color blindness is often thought of as a complete lack of color vision, where only black, white, and gray are seen, but monochromatic vision is exceptionally rare and requires malfunction of the entire cone system. In most cases, only one type of cone in the eye is partially compromised, damaged, non-functional, or absent. A protanomaly occurs when the red cones act abnormally or there is a partial loss of function, whereas protanopia indicates complete non-function of the red cones. Similarly, green cones can be impacted by a deuteranomaly or deuteranopia and blue cones by a tritanomaly or tritanopia. Because of the large overlap in wavelength ranges, malfunctions within the red and green cones are often collectively referred to as red–green color blindness. Protanopia, protanomaly, and deuteranopia impact about 1% of males each, and deuteranomaly impacts about 5% of males, resulting in about a total of about 8% of males who have some type of red-green

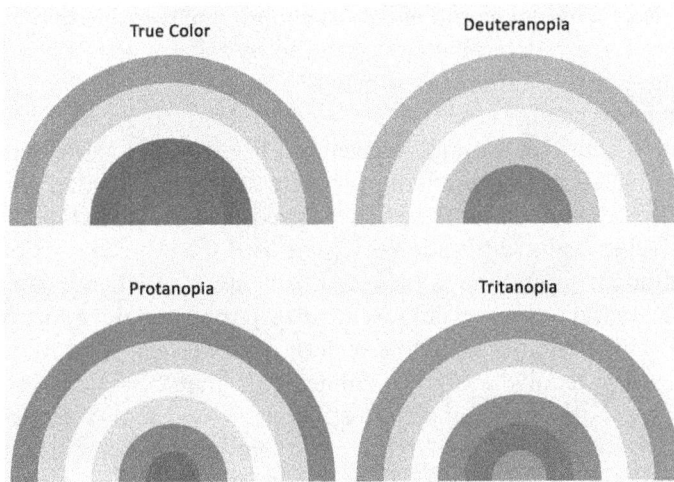

Figure 12.1. The top left image represents the colors of the visible light spectrum. The other images show a simulation of what the same image might look like to a person with an absent or fully malfunctioning cone type.

color vision deficiency. A common misconception is that color blindness is a disability that only impacts men, however about 0.4% of females have some form of color vision deficiency, the most common being deuteranomaly (0.35%). Tritanopia/tritanomaly is rarer, with estimates suggesting an impact of about 0.01% of the population for both males and females [4]. In figure 12.1 a simulated view of protanopia, deuteranopia, and tritanopia is provided as compared to full color image. The simulated image demonstrates how both protanopia and deuteranopia impact color vision in similar ways.

12.2.3 Other visual disabilities

Approximately 2.3% of the population has a visual disability that impacts their ability to interact with the world in a typical way, even with corrective lenses [5]. About one third of all visual impairments are caused by cataracts, where the lens of the eye becomes opaque and vision is blurred [3]. Some other common visual impairments like glaucoma (a condition of the optic nerve), retinopathy (particularly associated with diabetes), and eye damage caused by infection can cause loss of vision in sections of the visual field, difficulty focusing, loss of peripheral vision, loss of vision in the center of the visual field, tunnel vision, etc. All these conditions can result in difficulty interpreting graphics in presentations and publications, as well as difficulty recognizing signage.

12.3 General strategies for accommodating visual impairments

12.3.1 Color choice

Despite the facts that the colors red and green are the most difficult colors to differentiate for about 8% of the population, the use of these two colors for differentiation in images is often an industry standard. A wide variety of graphics traditionally use this color combination ranging from volcano plots in bioinformatics to the red–green diverging color scheme on relief maps. When possible, the use of red and green to differentiate between two different features of a graphic should always be avoided. Those with elevated professional status in their respective fields might even consider breaking convention in favor of a more accessible color scheme in a publication or presentation to help set a new standard for their field. Design professionals suggest combinations of green and purple in places where green and red color schemes have traditionally been used [6]. An example of this option with a deuteranopia simulation is provided in figure 12.2 where a simulation of a map using a diverging red–green color scheme (top row) and diverging purple–green (bottom row) simulated as it would be seen by a viewer with deuteranopia.

There are many readily available online tools that can aid you in choosing accessible color palates. A popular tool is the web-based tool ColorBrewer 2.0 [7]. While ColorBrewer is designed to provide color advice for cartographers, the webpage will provide a list of codes for color palates that are 'colorblind safe' for up to 11 data classes in either a sequential or diverging scheme. These color schemes can easily be applied to most applications. In addition, many users have developed

Figure 12.2. The top row of the image shows a map colored with a diverging red–green color scheme (left) and the map as if viewed by a person with deuteranomaly (right). The bottom show shows the same image colored with a purple–green scheme (left) and a similar deuteranomaly simulation applied (right) to demonstrate the power of a purple–green alternative to traditional red–green schemes.

plug-ins and code for color palates that are optimal for color deficient viewers in a variety of software packages including the popular open-source statistics software package, R. Packages that provide only color options that are ideal for audiences with color vision deficiencies can be found in online repositories for most popular plotting software.

12.3.2 Textures and patterns

A general rule that will help make all visualizations more accessible is that color should enhance graphics, but not be required to interpret them. This provides more benefits than just accessibility. For example, many people print publications, preferring to read hard copies, and often those copies are produced in black and white only. Tips for improving accessibility will also help with these types of practical concerns.

Plots of data are better interpreted when they use variety of patterns or shapes to convey differences as opposed to only being delineated by color. On scatter plots, different shape markers can be used as in figure 12.3. Trendlines should be adjusted to display in different patterns or line weights as opposed to just different colors. In the sample data set show in figure 12.4, different fill patterns are used to avoid issues with differentiation in color shade. Filling spaces with combinations of color and pattern is an ideal way to ensure that figures can be interpreted without color. Finally, because those who have some form of cone deficiency or malfunction can only see a reduced palate of colors, so the weight/thickness of lines and fonts will have a great impact on the ability of those with deficient color vision to differentiate colors because the shades of colors that they see are much more difficult to differentiate than shades of color seen by those with typical color vision. Lines with thicker weighting provide more color on the screen or paper and will be easier for those with color vision deficiencies to differentiate.

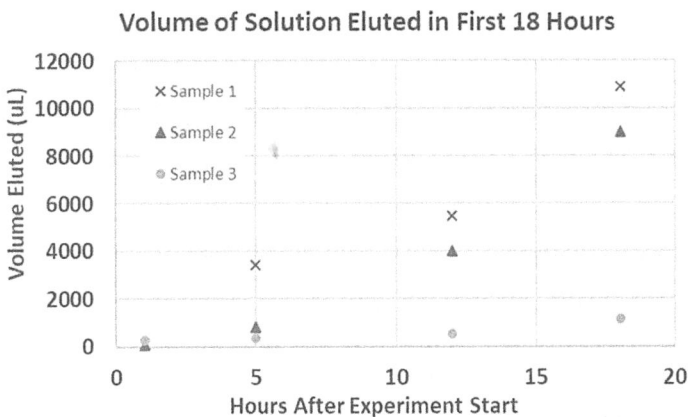

Figure 12.3. In this example data set, different samples are denoted by markers that not only use a more accessible color palate, but are also different shapes.

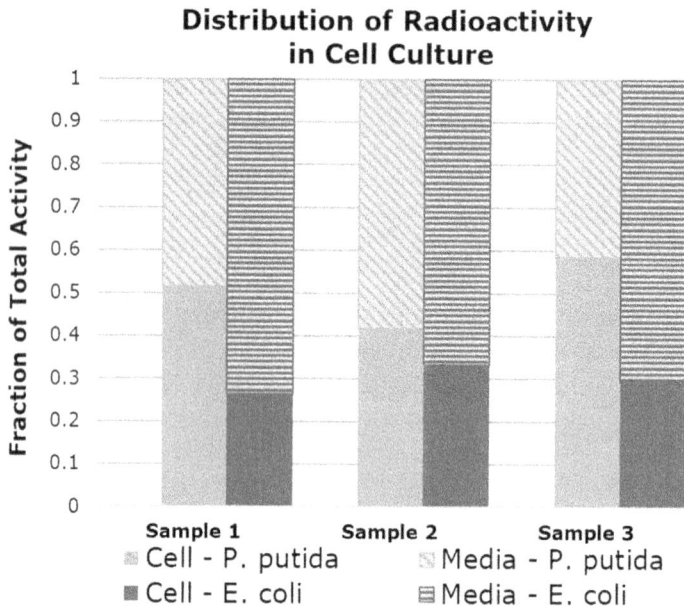

Figure 12.4. Stacked bar graphs, like this one from a sample data set, often use different shades of the same color to indicate that they are part of a whole sample. Because shading can be difficult for those with deficient color vision to differentiate, a pattern is used instead to help guide the reader.

12.4 Additional considerations

12.4.1 Presentations

The contrast of text and images in comparison to background is an important factor in how well an audience can see a presentation. There are some general recommendations to maximize accessibility of your presentation based on contrast ratio of text and line-based graphics. It is recommended that text has a contrast ratio of at least 4.5:1 but it is preferable to have a contrast ratio of 7:1. Thicker logos and images can have a contrast ratio of 3:1 [8]. If you are unsure how to check the contrast ratio of your presentations, there are many online tools available for this purpose. Charts and graphs should be no exception to these rules, where sufficiently-sized, bold fonts should be used to denote axis titles, axis values, and legends. Your audience will have difficulty understanding your data if they cannot read the axes on your plots.

Even with great attention to detail, a generally accessible presentation may still not be enough for an audience member with low or no vision. One way to ensure that your presentation reaches the greatest number of people is to take time to describe your graphs, charts, and other images with words. This might include describing trends or the location of data on a graph, using words to explain an experimental process depicted in photos, or any other descriptive process that would be useful in understanding the content of your presentation. Besides making your presentation more accessible to people with visual disabilities, this skill may prove useful in other situations such as meeting rooms with poor seating or projection quality or when presenting on conference calls where all participants do not have access to your presentation because of technical difficulties.

12.4.2 Publications

Ideal formats of color and text/pattern that will help make your work more accessible may not always meet the preferences of your publisher or may not have the visual appeal that those with typical vision prefer. In recent years publications have developed an increased capacity for supplements to publications via online resources. One option for charts, maps, or graphs that must meet a particular standard that reduces their accessibility is to include a more accessible version of the image, chart, or graph as part of a supplement.

12.4.3 Accessibility on the Web

Scientists, engineers, and safety professionals are sharing more information and data on the internet than ever before. It is not safe to assume that the way you see your webpage is how it will be read by a user with a visual disability. Many users with visual disabilities have plug-ins that invert colors, convert to grayscale, or increase contrast. Users with particularly low or no vision may use adaptive technology like a screen reader (a program that reads internet content ranging from web pages to social media) to them in an audio format. There are methods to include captioning (alternative text) for pictures on both social media and web pages that is compatible with screen reading technology. A website run by Toptal will even run a simulation of any webpage address entered with a filter for all four kinds of color deficient vision so that you can see how your website looks to all viewers [9].

The best way to ensure that your online content is accessible to most users is by following the accessibility criteria set by the World Wide Web Consortium (W3C). The guidance issued by the W3C is a comprehensive guide that provides standards for many aspects of accessibility and is not limited to visual disability. The W3C's web content accessibility guidelines are a great starting point for making web content accessible, and the standards are updated regularly to ensure that the guidance is both modern and compatible with current technology [8]. If you develop new web content with the aid of a professional design service, be sure to ask about how they can optimize your page from an accessibility standpoint.

12.4.4 Safety and visual impairment

Often, signage that indicates a warning of danger utilizes red, yellow, and green. As shown in earlier figures, these colors will not be easily differentiable to a large portion of people. While the use of color as a warning can be useful, be sure to never assume that color will be a sufficient indicator of danger. As with most internationally recognized signage standards that do not depend on color, all safety messaging should take this into account. Safety signage can also benefit greatly from considerations regarding contrast, as discussed in section 12.4.1. High contrast is an important part of making your signage as visible as possible. Finally, remember that there is no single solution to accommodating every visual disability. Take time to learn about the people who work in areas that you supervise, especially areas with hazardous conditions, and find out if there are employees with visual disabilities that might require accommodation. Asking an employee with a visual disability how they can best be accommodated is the most efficient and safest way to ensure that appropriate messaging is provided and maintain a safe working environment.

References

[1] Hendricks A 2007 UN convention on the rights of persons with disabilities *Eur. J. Health Law* **14** 273

[2] Brault M W 2012 *Americans with Disabilities: 2010* (Washington, DC: US Department of Commerce, Economics and Statistics Administration, US Census Bureau)

[3] World Health Organization *Global Data on Visual Impairments 2010* WHO/NMH/PBD/ 12.01.

[4] Snowden R, Thompson P and Troscianko T 2006 *Basic Vision An Introduction to Visual Perception* (Oxford: Oxford University Press) pp 382

[5] National Federation of the Blind 2018 *Blindness Statistics [Internet]* [cited 2018 Nov 1] Available from: National Federation of the Blind

[6] Jenny B and Kelso N V 2007 Color design for the color vision impaired *Cartogr. Perspect.* **58** 61–7

[7] Brewer C, Harrower M, Sheesley B, Woodruff A and Heyman D 2013 *ColorBrewer: Color Advice for Maps [Internet]* ColorBrewer 2.0 Available from: http://colorbrewer2.org

[8] Web Content Accessibility Guidelines *W3C Recommendations [Internet]* W3C Guildlines (05 June 2018) Available from: https://w3.org/standards/webdesign/accessibility

[9] Toptal *Toptol Color Blind Filter [Internet]* Available from: https://toptal.com/designers/ colorfilter

Part III

Not the usual topics, going outside
the space–time continuum

IOP Publishing

Laser Safety
Practical knowledge and solutions
Ken Barat

Chapter 13

Ergonomics in a laser lab, you must be joking

Ken Barat

13.1 Introduction

When an employee is assigned a desk in an office setting it is not uncommon for that individual to receive an ergonomic evaluation. The goal is to prevent carpal tunnel injuries that would reduce their productivity and increase company cost. This activity and cost are just part of doing good business. But when an individual is assigned to work in a laboratory setting that mind-set seems to be missing.

One rationale given is that little time is spent in the lab at a workstation; for myself, I do not accept this excuse or rationale.

The three basic ergonomic factors in the lab are posture/position, repetition/duration and force. Remember your Mom saying sit up straight?

13.2 Lab ergonomics, reaching across the table

The most common issue in the laser lab relates to reaching for optics equipment and the strain it can place on one's body. Since one never has enough room in your laser workspace, being able to reach all optics etc can become a challenge. Many times, one cannot arrange the lab so that access to the optical table is possible from all sides. In some cases, pieces of equipment become the obstacle that makes accessibility difficult.

Depending on how often one must reach for these optics, it might not seem like a concern. Please keep in mind that one minor torqueing of your back or body may be the start of a lifetime of pain and discomfort.

What are the solutions? Each of the following deserve some consideration, they will not be the answer to all set ups, but are worth considering.

1. *Motorized mounts*

    Such mounts are controlled off the table and make reach in unnecessary and should be partnered with remote viewing cameras.

2. *Remote viewing with cameras*

The cost of these items has decreased so that their use must be considered. Their use removes the user from harm's way in many set ups and can give multiple views depending on how they are set up and displayed. For example, looking at several security cameras on a screen.

3. *Platforms to stand on giving greater reach capability*

These come in different heights, just like staff, and various commercial platforms are available.

4. *Vertical breadboards on table*

At times these can bring optics closer to the reach of users and save space.

5. *Set up so one can reach in from all sides or at least three*

This is just good planning, take advantage of building codes to get extra room.

6. *Raising some items on shelves*

While this means elevating some beams, the risk might be balanced by increased accessibility.

7. *Optical table modifications*

Some optical tables can be configured with a hole in the middle, like model railway tables for access, and still maintain optical table properties. While I have rarely seen anyone use this option, this might be worth considering in initial lab setup.

8. *Classical ergonomics*

Lastly is standard ergonomics for the lab; the height and position of monitors, where keyboards are placed, the type of mouse used etc. These items are the same as routine office ergonomic safety but are rarely considered in lab set ups. The usual reason given is 'how little time a person spends working with them'. To me that is like saying 'I do not use my seat belt on short trips'.

Classical ergonomics in a lab setting needs to be looked at by someone familiar with ergonomics. The typical rules are:
   - Monitors should be directly in front of you with the upper edge at eye height or slightly below; needs to be adjustable.
   - Use a document holder for hardcopies and keep in front of you, between monitor and keyboard.
   - Keep the keyboard and mouse in front of you and as close as practical to prevent over reaching.
   - Wrist straight as possible.

9. *Trip hazards*

I just want to add that trip hazards need to be addressed. This can be done by 'bridges' designed to contain wires and hoses.

The Occupational Safety and Health Administration provides a review and recommendations through its eTool on ergonomics.

Work in the laser lab: During initial set up of the optical laser path, experimenters are required to install the optical components onto the laser table. The typical optical table is 3×6 feet and 4×20 is not uncommon.

13.3 Typical work activities and posture

The experimenter is required to stoop and reach over the table at various distances depending on the position of each optic. This duty cycle has a range of 30 minutes to several hours. The duty cycle is defined as the length of time needed to complete the task of fully aligning the beam on the complete table. Once the table is set up, alignment must be initially and periodically performed. This task typically involves stooping over the table and manually adjusting optics in three dimensions with either manually adjustable knurled finger screws or with a small screwdriver. Stooping is defined as bending at the waist while maintaining extension of the lower extremities at the knees. Forces required by the hand are minimal. Each adjustment has a cycle time of 30 seconds to a minute and can be required for one to tens of optics. The cycle time is defined as the time necessary to align one optical device on the laser table. Reaching to the center of the table in limited access areas, may require a stretch reach from a stooped posture. After daily alignment, the researchers may then spend part of their time at a computer station remote from the laser table or at the optical table edge. Re-alignment varies between researchers from routine daily alignment of each optic to minimal alignment every few weeks.

13.3.1 Do you have back pain?

Researchers who use lasers need to repetitively stoop over the light table in order to set up, initially align and to re-align laser optics. The repetitive stoop has been reported by workers as a source of lower back pain during interview. Symptoms have been described as lower back soreness after work and as 'tightness' in the lower back. Workers find it necessary to pace the alignment activities and the alignments are broken down into sections with periods of rest taken as needed prior to resumption of alignment activity. There is no published study regarding musculoskeletal injuries among laser researchers found in reviewing the PubMed database that was conducted in 2011.

Figures 13.1 and 13.2 show a sample of the workstations found in typical laser laboratories this author has visited.

13.4 Getting higher

Platforms that raise the user's height is one solution many have used to reach across optical tables. The drawback is that they take up floor space, making it harder to hide things under the optical tables. But being higher makes it less likely one will bump an optic when reaching across, unless their waist bulges down. Some examples follow (figure 13.3).

13.5 Standing around

Anti-fatigue mats, which are common in the food industry where people may stand for long periods of time, can also be an aid to researchers.

Figure 13.1. Keyboard examples.

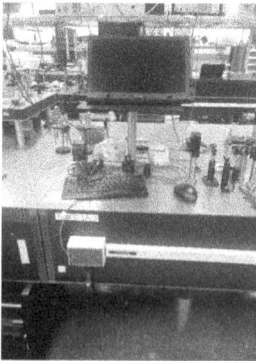

Figure 13.2. Workstation examples.

Figure 13.3. Platform examples.

Figure 13.4. Panels around an optical table made to be removed for access.

13.6 Weight/lifting

The lifting of pumps such as vacuum pumps (turbo etc) can easily cause back strain. If panels have been placed around an optical table setup which are designed to be removed (figure 13.4), consider their weight and how this action should be performed. Think ahead, not just considering the present staff or the 6 ft grad student, but that new, possibly just 5′2″ person. When I say lifting, it is not just vertical panels, but also horizontal panels that may be part of an enclosure of the beam path or laser system optics.

You just must do better than this (figures 13.5 and 13.6).

13.7 Vertical breadboard

Like driving on the 'wrong' side of the road when traveling abroad, it takes some getting used to, but has definite space advantages (figure 13.7).

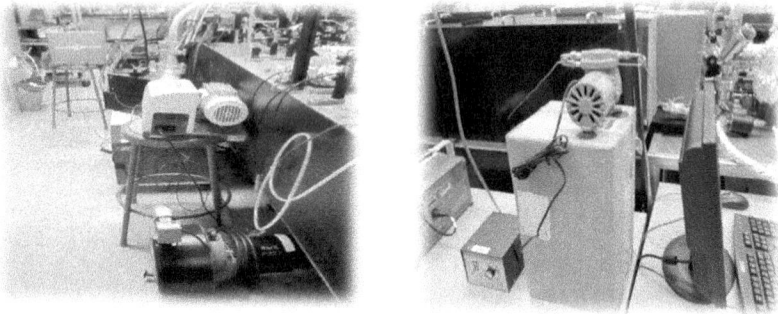

Figure 13.5. Unstable positioning of pump.

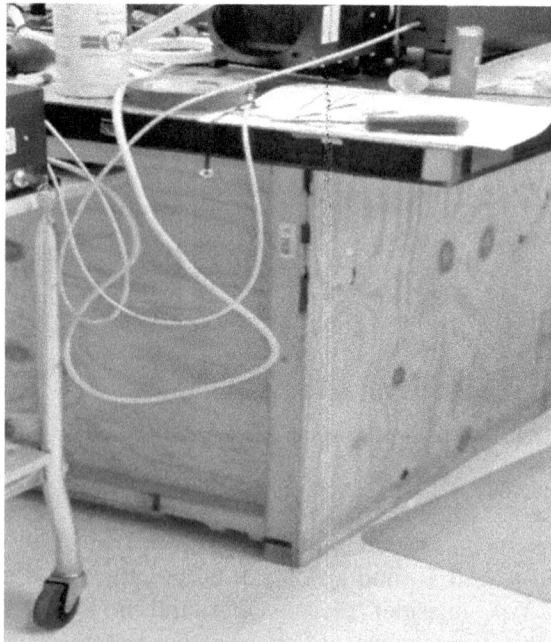

Figure 13.6. Poor breadboard support.

Figure 13.7. Examples of vertical breadboard usage.

13.8 Work hours

While not a typical ergonomic consideration, please remember excessive work hours are not a badge of honor or standard activity. Excessive work hours can, and often do, lead to bad decisions that have led to several personnel and equipment related accidents.

13.9 Concluding thoughts

Typical potential ergonomic hazards in a laser laboratory:
- Reaching across optical tables.
- Standing for long periods.
- Make-shift workstations with related carpal tunnel concerns.

Now that these have been brought to your attention, look around your workspace and see how they can be addressed.

Chapter 14

Laser safety tools: making your life better for less

Ken Barat

14.1 Introduction

When one thinks 'laser safety' most often it is laser protective eyewear that comes to mind. The LSO should know there are more laser safety tools than just eyewear. When a commercial product is named specifically it will be because, to my knowledge, that product is unique, otherwise generic product terms will be used. There are more products than this chapter can show, but the hope is that what is here will make you think of other things.

The following sections describe items that might be useful, depending on user needs.

14.2 Carbon resin lightweight breadboards

If weight is a consideration, carbon resin breadboards should be considered. They are more expensive that the standard metal breadboard but weigh 75% less. Resin breadboards are only available from European sources and hence the hole threads are not the typical size found in the US.

14.3 Vertical breadboards

While we are talking about breadboards, what about vertical breadboards (figure 14.1)? Like driving on the wrong side of the road, users can adapt to this approach. It does save real estate. They can be useful on a horizontal table as a space for pick off diagnostics. The alternative is to place the entire optical table on its side. Of course, then there is gravity to help housekeeping. Getting too close to optics with IR viewers can be a problem.

14.4 Black AL foil

Several sources to obtain these are on the web. Since they are non-reflective, inexpensive and can be bent into shape for temporary use, they are great as a beam

Figure 14.1. Vertical breadboards.

tube wrap to cut diffuse reflections and will hold their shape if bent over an open beam area etc.

14.5 Diffuse reflection material

Spectralon® diffusion material gives the highest diffuse reflectance of any known material or coating over the UV–vis–NIR region of the spectrum. The reflectance is generally >99% over a range from 400 nm to 1500 nm and >95% from 250 nm to 2500 nm. The material is also highly Lambertian at wavelengths from 0.257 mm to 10.6 mm, although the material exhibits much lower reflectance at 10.6 mm due to absorbance by the resin. The surface and immediate subsurface of Spectralon® exhibits highly Lambertian behavior. The porous network of thermoplastic provides multiple reflections in the first few tenths of a millimeter of Spectralon®.

14.6 Indirect laser beam viewing tools

14.6.1 Laminated IR viewing cards

The IR viewing card is designed to allow one to see invisible infrared beams. The majority of IR cards found in laser labs to protect the fluorescent material from oxidation are covered with a plastic film. Unfortunately, this yields a specular reflector, they are often held by hand (and hence wobbling at all angles). One suggestion is to peel off the coating or use non-laminated ones. Sensor cards can also be found for ultra violet wavelengths, but are less commonly used. *It is important to*

ALWAYS tilt the IR sensor card DOWN so that any reflection is not directed to yourself or anyone else standing nearby.

Sensor cards are not invincible, know your expected irradiance, you can burn through these cards and present a fire hazard. As a general rule, *never* leave an IR sensor card nor any combustible card/plastic/beam blocks in a beam path unsupervised for an extended period of time.

14.6.2 IR viewers

IR viewers have been a staple in laser labs for decades. The safety concerns with these fall into two camps. First, can one use it with their laser protective eyewear on? Depending on your eyewear the greenish view through a viewer may make viewing difficult, prompting removal of eyewear, yielding unprotected eyes.

Second, one is tempted to look at the beam directly, thinking of the viewer as being eyewear. Although a direct beam will not transmit through an IR viewer, a direct beam viewing through an IR viewer will likely have a blinding effect to the eye by overwhelming the sensor, as well as risking damaging the IR viewer.

Neither of these is good or safe practice.

A superior, though more laborious alternative, is remote viewing with an IR camera, which removes you from standing in front of the beam or reflection.

14.6.3 Hands-free IR viewer

The best of these are really designed as night vision systems and will not claim to be laser detection devices. Many will detect out to 1200 nm. Please note that as with any optics, superior optics will cost more. A $2000 USD pair of night vision goggles will give better resolution than a $400 USD pair.

A number of commonly used items, if not used properly, can become sources of hazardous reflections or dangers to the user (figure 14.2).

- REFLECTIVE ALIGNMENT TOOLS: *Beware to ALWAYS consider reflected beams off hemostats and IR cards during alignments.*

14.6.4 CCD/webcam

The webcam used in some smart phones, such as earlier model i-Phones can be used to view visible and NIR beams. A CCD camera is a superior option and can be placed on optical posts. The advantage of these devices is they remove one from the optical table. They come in a number of varieties, commercial to home-made. Combined with the use of motorized mounts this can make alignment a simple activity (figures 14.3 and 14.4).

Figure 14.2. Reflective objects.

Figure 14.3. CCD camera, remote viewing.

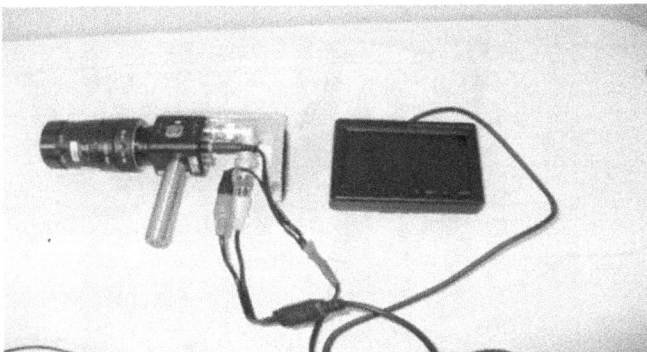

Figure 14.4. Home-made system.

14.7 Beam blocks

Many items fall under the definition of a beam block, not all were designed to be such. Use of a note card, post-it note or other paper based items as a beam block is not recommended (figures 14.5 and 14.6).

While inexpensive, they easily fall over, yielding a suddenly unblocked beam. They can also slowly (or not so slowly) burn through if placed at a point where the beam is intense enough, suddenly letting the beam through and maybe extending far down the optical table or off the table. They are commonly found used as blocks for optics transmission, block diffuse reflections or primary beams.

When using cards or paper as *temporary* laser shielding, it is important/essential to know which color to choose to avoid laser beam absorption in the card/shield and therefore, risk of burning and/or heating. Also note that leaving a card as a block in front of an optic will often outgas and leave residue on the optics which can ultimately damage it if it is not cleaned properly.

14.7.1 Unsecured beam blocks

The majority of beam blocks (metal) are designed to be secured to the optical table, either by being screwed down, having a magnetic base or just by their weight and center of gravity. With beam blocks that cannot be secured, because they might need to be easily moved, they can also be accidentally easily moved out of position or knocked down. This type is usually bent sheet metal or folded cardboard. The range of size and protection level of beam blocks vary.

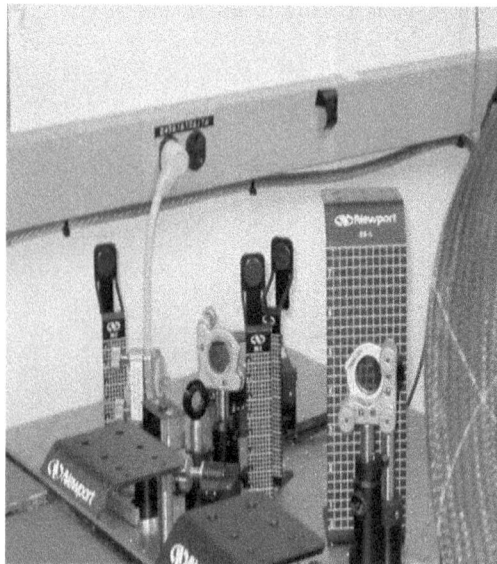

Figure 14.5. Commercial beam blocks varying width.

Figure 14.6. Paper beam block.

Figure 14.7. Home-made beam blocks.

14.7.2 Home-made beam blocks

The fact that standard beam blocks are 1/8 to 1/4 inch metal 'L-shaped' (figure 14.7) means that any facility with a metal shop can make these. It is best if they are anodized or beam blasted to give a diffuse surface. It is often a matter of the cost and time, making them yourselves costs less than buying commercial ones. Paper, post-it notes and other materials, while commonly seen as beam blocks, are risky (figures 14.8 and 14.9).

14-6

Figure 14.8. Tape used as a beam block.

Figure 14.9. Paper beam blocks.

14.8 Beam dump

A beam dump can be considered a heat sink. It captures a diverted beam. These are either air or water cooled depending on the amount of energy they are intended to deal with.

Pipe fittings can be used if you have no budget (figure 14.10).

14.9 Polycarbonate sheets

These can be used as perimeter guards for UV and carbon dioxide wavelengths. Thus giving a clear view of the optics on the table.

14.10 Plastic laser enclosures

Plastic/acrylic laser enclosures that are rated for certain wavelengths and provide a tested optical density (filtration) can be expensive (figure 14.11). Most commonly people buy plastic or acrylic sheets from a supply catalog. Depending on the wavelengths being used they provide an effective containment for scatter or direct beams. One of the better designs is having a diffuse film on one side (should be set as the interior not exterior surface of the enclosure). Using a spectrometer and/or power meter, one can self-test the materials. The choice of a proper plastic laser enclosure should never be based just on a visual or 'feel good' evaluation. Remember as one adds new wavelengths to a system, the enclosure's containment should be reconfirmed.

14.11 Metal laser enclosures, table perimeter guards

These would seem to answer the uncertainties listed above for plastic enclosures. The cautions with metal enclosures is burning off the coatings, and making sure it does not present a specular reflection source (figure 14.12).

Newer styles allow cables and tubes to leave the table while still providing a high level of protection (not shown here, image permission problem).

14.12 Laser curtains

Laser curtains are most commonly used to segregate areas of a laser lab. At ceiling height, they interfere with the water distribution pattern of fire suppression

Figure 14.10. Pipe fitting used a beam dump.

Figure 14.11. Acrylic and Lexan enclosures.

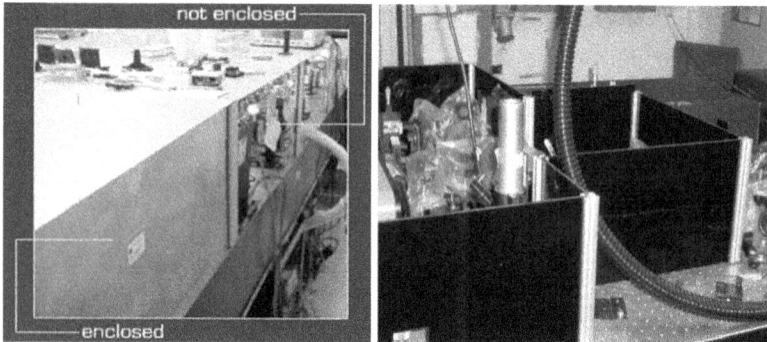

Figure 14.12. An approach on cable pass through.

sprinklers, meaning that the sprinkler heads need to be lowered to be effective. Laser curtains can be certified laser curtains or in some cases opaque welding curtains, contact the LSO for options. There is a considerable price and performance difference between the two. Laser curtains can also be made of metal (figure 14.13). See chapter 15 for more on curtains.

Figure 14.13. Curtain use example.

Figure 14.14. Use of pipe insulation as bumper guards.

14.13 Laser protective eyewear

Laser protective eyewear is one's last line of defense against laser beam exposure.

There are more things out there that can have laser safety applications, those mentioned in this chapter are just to get you thinking. Kentek, Thor Labs, Lasermet, Newport and Laservision are all good sources for items that can help make the laser user safer.

14.14 Piping insulation as shelf/head guard

If you have shelves hanging over your optical table you have a built-in head knocker. An easy solution for such shelves is the inexpensive padding used for pipe insulation (figure 14.14).

14.15 Are there more products out there?

The answer is yes, many are specialized to applications, others are broader. The lesson learned is always keep your eyes open for items to apply to your laser safety needs.

IOP Publishing

Laser Safety
Practical knowledge and solutions
Ken Barat

Chapter 15

Evaluation and design of laser barriers

Tom MacMullin

15.1 Introduction

From our extensive relationships working with customers in many countries and associated with almost every imaginable laser application, we are acutely aware that even the most experienced laser person typically has a very narrow understanding of laser barriers. The purpose of this chapter is to continue building within the laser community a shared understanding and appreciation of laser barriers.

We have put these notes together in a way to make them accessible to laser users of every experience level. All terminology will be explained; synonymous usage of relevant terms is highlighted. We hope the novice laser person and non-laser designers of laser laboratories will be informed by our experience. We believe that the most experienced laser users will also find some useful insights and knowledge in this text.

The mandate for this chapter is derived in part from ANSI Z136.1, the *American National Standard for Safe Use of Lasers*. Laser Control Measures as defined in section 4 of that standard require that no one be exposed to unsafe laser radiation, defined therein as radiation levels that exceed maximum permissible exposure (MPE). Relevant excerpts from ANSI Z136.1 are found in the nearby table 15.1. Similar conditions and safety requirements are also described in other standards that apply in the European Union and countries that follow EU regulations. See table 15.2 for a list of the most relevant standards that apply to laser barriers, screens and curtains.

The ANSI Z136.1 standard also requires access controls into laboratories and other spaces where Class 3B or Class 4 laser hazards may be present. The three types of access controls commonly used, and often found in conjunction with the laser barriers that are the main topic of this chapter, are:

    (i) Entryway interlocks
    (ii) Mechanical or electronic door locks
    (iii) Warning signs/lighted signs

doi:10.1088/2053-2563/ab0f25ch15   15-1

Table 15.1. Laser barrier requirements from ANSI Z136.1.

Section	Title	Summary of guideline
2.2 Definitions	LCA	Laser controlled area is a laser use area… (which) may be defined by walls, barriers, or other means. Within this area, potentially hazardous beam exposure is possible.
2.2 Definitions	NHZ	Nominal hazard zone is the space within which the level of the direct, reflected, or scattered radiation may exceed the applicable MPE.
2.2 Definitions	MPE	Maximum permissible exposure is the level of laser radiation to which an unprotected person may be exposed without adverse biological changes in the eye or skin.
4.4.2.5	Laser Protective Barriers and Curtains (Class 3B or Class 4)	A blocking barrier, screen, or curtain that can block or filter the laser beam at the entryway should be used inside the laser controlled area to prevent the laser radiation from exiting the area above the applicable MPE.
4.4.4.1	Personal Protective Equipment	Enclosure of the laser equipment or beam path is the preferred method of control since the enclosure will isolate or minimize the hazard.
D3.4	Appendix—laser barriers and protective curtains	Area control can be effected in some cases using special barriers….designed to withstand either direct and/or diffusely scattered beams.

Laser users looking for more information about how to integrate laser safety tools in addition to laser barriers should consult the SPIE publication *Laser Safety in the Lab* authored by Ken Barat.

15.2 Laser barrier definition

A laser barrier is a device used to block or attenuate incident direct or diffuse laser radiation. Laser barriers can be permanent, such as hanging curtains and optical table surrounds, or may be temporary as exemplified by a portable barrier set up to protect nearby non-laser personnel during laser maintenance procedures. This definition is broad, and the range of equipment and materials available to serve as laser barriers is extensive.

Table 15.2. Important standards for laser barriers and screens.

Standard	Title	Notes
ANSI Z136.1	Safe use of lasers	As the parent document of the Z136 series of laser safety standards, the Z136.1 is the foundation of laser safety programs for industry, military, research and development (labs), and higher education (universities).
EN 12254	Screens for laser working places	This standard specifies functional requirements and product labeling applicable to temporary and permanent passive guards (called screens or barriers) for protection against laser radiation.
EN 60825-4	Safety of laser products—Part 4: Laser guards	Specifies the requirements for laser guards, permanent and temporary (for example for service), that enclose the process zone of a laser processing machine, and specifications for proprietary laser guards.
NFPA 701	Standard methods of fire tests for flame propagation of textiles and films	Fabrics used in most public spaces are required by law in most states and cities to be certified as flame retardant. This test measures the flammability of a fabric when it is exposed to specific sources of ignition.
ASTM E84	Standard test method for surface burning characteristics of building materials	The flame spread index and smoke developed index values obtained by the ASTM E84 test are used by code officials and regulatory agencies in the acceptance of interior finish materials for various applications.

Laser safety control measures include administrative controls such as procedures, training, warning signs, and personal protection, and engineering controls which may include protective housings, interlocks, beam stops, barriers and curtains. The laser safety rules and guidelines established by the ANSI Z136.1 standard are interpreted to prefer engineering controls in a laser environment, especially when administrative controls with similar risk-reduction results are being considered.

Any student of trigonometry would determine that laser barriers are best placed as close as possible to the laser to provide extended angular coverage. Thus, engineering controls can be viewed in layers. The laser itself is ideally placed in a box or enclosure to create Class 1 conditions. Next, the optical table or production work cell may have integrated laser barriers, possibly defining a perimeter. Finally, the walls and doors can be considered in the overall barrier system. All these constructions need to be evaluated as a barrier against the Z136.1 standard.

Before we embark on a detailed investigation of laser barrier curtains and laser barrier partitions, we should mention some of the equipment found in the 'inner layers' of the laser environment that qualify technically as laser barriers. Many lasers are equipped with a shutter that includes an integrated laser radiation absorber or beam dump. Energy deflected by a shutter or redirected by other optics in a laser setup may terminate in a stand-alone beam dump, beam trap or beam block. Also recognize that power meters sometimes serve as a barrier when, for example on a university optical table, a fixed portion of a laser beam is continually directed into a power detector for monitoring purposes. Beam dumps as part of a shutter, integral to a power detector, or mounted stand-alone may be air-cooled or water-cooled.

15.3 Laser barriers–curtain style

15.3.1 Overview—laser barrier curtains

Laser barrier curtains are comprised of a select barrier material, suspended from a track system, and operated with rollers or slides that guide the barrier into place. It follows that the positioning of such curtains is fixed, and the areas that curtains delineate are also fixed. A corollary is that laser curtains only work as proper barriers when they are moved into place. A conclusion that must be drawn is that the laser barrier curtain, as an engineering control, likely depends on other engineering controls, such as interlocks, or administrative controls, such as a procedure that includes drawing the curtain into position.

Laser curtains are used in research, industrial and medical facilities to separate eyewear versus non-eyewear zones. If the laser hazard is contained within the Laser Controlled Area (LCA, see ANSI Z136.1), then the curtain establishes a laser exclusion zone or laser free area (see Barat, op.cit.). Optionally, a laser barrier curtain may separate multiple lasers, each with its own unique radiation hazard. An example of this is a large room in a university science complex in which multiple optical tables are present, each with a different laser. In this situation, a system of laser curtains is installed to segregate the optical tables.

15.3.2 Laser barrier curtain materials

Laser barrier curtains are constructed from a wide variety of materials. Laser users should make intentional, not subjective, selection of material(s). The barriers should be tested in accordance with relevant standards, EN 12254 for example, plus standards related to local or national building codes such as NFPA 701 and similar. The barrier materials should be clearly marked with irradiance values and other performance metrics that enable selection of barriers by laser personnel.

Plastic laser curtains: Plastic curtains may be opaque, translucent, or clear. When used in a laser environment, plastic barriers are generally specified and selected based on optical density properties and in this regard may be evaluated with the same care with which laser safety eyewear is chosen based on the ANSI Z136.1 standard. Opaque plastic curtains may block diffuse and low energy radiation but plastics tend to have lower damage thresholds when compared to other barrier materials. Included in the category of plastic curtains are light and heavy gauge films and possibly welding screen materials.

Fabric laser curtains: Fabric curtains are constructed of natural and/or man-made fibers processed to create opaque substrates. The fabrics may be woven, coated, laminated, or layered by mechanical attachment including stitching. Some laser barrier fabrics are comprised of multiple layers while others are single layer or monolithic. A few multi-layer fabrics feature a 'laser facing side' and an opposing 'visitor side'.

Metal laser curtains: Curtains made entirely of metals that work on conventional track systems are also available. The lightweight metal panels offer very high laser damage thresholds. These systems have many advantages over their fabric-based cousins yet do not cost significantly more. Choose a metal curtain, for example, if there is a possibility that laser radiation within the hazard zone could increase significantly over time as the complexity of experiments increases or maybe as a change in production equipment occurs.

Figure 15.1 shows a straight run of metal laser barrier curtain with a valance, an infrared laser viewing window, and interlocks on the entry door. Note the floor gap left open to facilitate both cleaning and air flow.

Multi-layer laser curtains: Multi-layer 'blanket style' curtains combine the pliability of fabrics and the power density ratings of metal by creating a sandwich of aluminum film or other metallic layers between outer layers of laser barrier fabrics. These designs significantly boost the damage threshold ratings of the laser barrier at the expense of thickness and bulk. Specifically, these barriers consume more floor space when they are stacked and put away at the end of the track system.

Figure 15.1. Metal laser barrier curtain.

15.4 Barrier selection

Choosing a laser barrier curtain is informed by multiple, and sometimes opposing, requirements and constraints. Foremost among these factors is the overall laser protection level. Using ANSI Z136.1 as a guide, use this process as a guideline:

- Step 1: Determine the nominal hazard zone (NHZ).
- Step 2: Determine the required damage threshold level for the barrier.
- Step 3: Select the material that is highly absorbing, will survive the laser threat, and satisfy ANSI Z136.1 taking into consideration all (!) of the factors described in the next few paragraphs.

Wavelength dependency: Do you need optical density type of protection or is opaque damage protection a requirement? Does the material provide the coverage you need for your wavelength? We have found some laser barrier materials that have protection levels which differ according to wavelength.

Permeability: Some fabric laser barrier materials are woven too loosely to provide safety from laser radiation. These fabrics, too, may not be suitable for creating low-light or blackout laser room conditions.

Formability: Will the material allow creation of shapes and configurations you need to cover your environment? Does the barrier material, in its desired configuration, unreasonably constrict the workspace? Pay attention to how a curtain stacks when it is moved into its full storage position. Review whether your laser barrier curtain needs valances at track level and decide how those will be constructed.

Accessories: Will the laser barrier material, in its desired form, work with the other items your laser environment needs, including possibly doors, windows, interlocks and other laser safety accessories?

Particulates: Review whether the laser barrier curtain materials create unwanted particulation or outgassing, both in standing mode over time and when there is a laser-induced particulate release. Note that silicone coatings are not acceptable in very sensitive environments due to potential outgassing concerns. Do you expect your laser barrier curtain to be installed in a clean room, and if so, is special treatment or preparation of the laser curtain required before installation?

Life safety: Does the barrier satisfy applicable standards for fire and/or flame retardance? Some states, and certain municipalities, impose testing and certification requirements that exceed the general national standards. The barrier at a minimum should be clearly marked for relevant flame retardancy characteristics.

Other environmental needs: Are biohazards likely to be present in the laser environment and, if so, will the barrier material tolerate cleaning and/or disinfecting on a periodic basis? Is the laser barrier expected to serve as a physical security barrier? Does the laser barrier in its desired form need to provide blackout or low-light conditions in the laser space?

Aesthetics: Will the laser barrier be attractive, will it be visibly pleasing? How important is color to your environment? Black barriers generally absorb more laser radiation, especially in the visible range. White barriers are believed to reduce risk by minimizing dilation of pupils. Black, shades of black, white, gray, and blue laser barrier curtains are all available. Does surface texture matter? Consider that smoother surfaces are easier to clean, but also may enhance reflection of laser radiation depending on wavelength.

15.5 Laser barrier curtain design

The overall design of the laser barrier curtain system must create the desired laser free zone without compromising other aspects of safety and impart minimal impact on the usability of the entire laser room or production cell. It takes technical know-how, deep experience, and an occasional burst of creativity to specify a curtain system that laser technicians will use to preserve levels of safety without being tempted to override the controls put in place. The system must be both laser safe and user-friendly.

The ceiling: It is tempting to design the overall curtain track configuration around just the production equipment or adjacent to the optical tables. We encourage laser room designers to spend significant time looking at the ceiling: most room ceilings present multiple obstacles that will limit track runs or require special turns and intersections. Lighting, HVAC machinery, beams and ceiling supports all create constraints on the final layout.

Figure 15.2 shows a curved laser curtain track segment attached to both a cluttered ceiling and a mid-room support column.

Height: Laser barrier curtain heights are often limited by existing ceiling heights, but where the installer has flexibility, there are several factors to consider. Fire suppression equipment or the requirements for heating, cooling and ventilation may prevent the curtain system from being extended to the maximum possible height. To enhance laser safety when the laser curtain creates a boundary inside the nominal hazard zone and when faced with height restrictions, ensure that potential sources of specular reflection overhead are minimized. Also, it may be necessary to move the curtain perimeter either toward or away from the laser hazard per the

Figure 15.2. Curtain support from ceiling and posts.

recommendation of the site Laser Safety Officer (LSO) who can help maximize efficacy of the laser barrier.

Valance: Laser barrier curtains are typically suspended from a track system using a system of rollers or slides with hooks or similar fasteners that extend downward from the track. There is a built-in gap for potential laser excursion between the bottom plane of the track system and the top of the barrier materials. Best-in-class laser curtains are manufactured to ensure the smallest possible gap at the top of the barrier system. In situations where this gap creates an unacceptable weakness in laser safety, a valance or shroud of laser barrier material may be installed at track or ceiling height, on one or both sides of the track, either to create a beam trap or to establish a continuous laser barrier that includes the entire desired height.

It is our experience that many valance installations reflect over-design from a laser safety perspective but, if properly constructed, add a significant aesthetic boost to the laser barrier. Some valances may impede the operation of the curtain, particularly when a curtain is pulled open and the fabric stacks up accordion-style such that it drags along the face of the valance. An option to valances is to install the curtain track in a recessed channel (a soffit) built into the ceiling. If this option is taken, ensure that the construction process accommodates the need to feed rollers into the track and to hang the barrier curtain onto the rollers. We have encountered several beautiful new buildings with decorative soffits that had to be partially dismantled to finish the curtain install.

The floor: How low should your curtain be? For heating and air conditioning, it may be best to leave a short gap at the bottom of the curtain, typically from one-half to two inches in height. This also helps with maintenance including floor cleaning. There are several possible situations for which the barrier curtain should drag on the floor: (1) concern over impact of potential floor reflections on laser safety; (2) desire to create a blackout or low-light installation; or (3) need for limiting air flow around the laser safety zone. Our experience is that many floors are not completely level and desired floor height or drag height may not be achieved if dimensional tolerances are not carefully evaluated. The choice to leave a gap or create a floor drag also depends on the access or egress points; some laser barrier curtain designs feature doors built with rigid materials that need a specified minimum floor clearance to operate properly.

Two other aspects of the interaction between a floor and a laser curtain may be important to some users. First, depending on the laser application and the laser barrier material chosen, it may be necessary to ground the curtain for static discharge. Some curtain manufacturers offer optional static collection strips embedded in the product with contact points for connection to room or building ground systems. Second, the bottom hem may need to be weighted to ensure a seal against laser radiation, prevent leakage of ambient light into the laser area, or simply to hold the curtain down to reduce billowing or movement in the barrier wall induced by vehicular or pedestrian traffic. For example, we are aware of at least one installation of a laser barrier that created a long curtain wall that was immediately adjacent to a fork lift aisle in a manufacturing facility. Each pass of the fork truck caused the curtain to move enough to both physically disrupt the laser work area and cause chattering of the interlocks on the curtain.

Access and egress: Access points for personnel to move through the laser barrier curtain system are critical to safe design. Entry into a controlled laser zone should occur at a place that minimizes potential laser risks. Exit from the laser area must enable quick and unobstructed egress in the event of an emergency. Laser laboratories tend to be dark environments; egress points should be easy to find and operate with low or no light. Large rooms with multiple lasers segregated by laser barrier curtains need well-coordinated doors or access points to ensure that users in one section do not inadvertently create an unexpected laser hazard for personnel in an adjoining bay.

Zippers or hook-and-loop (Velcro®, e.g.) fasteners at curtain joints are efficient to manufacture and familiar to operate but may not be the most laser safe choice. Few university students will take the time close the zipper, and only an elite few are tall enough to reach the curtain top to re-seat the hook-and-loop closures. The barrier is not effective and a hazard exists when the barrier is not fully closed.

An interlocked entry point into a laser work area is often desirable. Many interlock switches use a magnetic reed or proximity switch. Fabric laser barrier curtains are difficult to interlock properly because movement in the flexible barrier at the point of egress may cause the switch to chatter or not engage. This is solved by adding rigid materials, such as strips of metal, to the opposing surfaces of the curtain to create a more formal doorway.

When interlocks are not required, we recommend a curtain entry point that features a trap for the laser radiation that is opened only when the curtain is opened. This trap can be created by designing an overlap of panel sections or a covering flap of barrier material—egress is achieved by pushing through an entry that closes behind the user. Such a trap is demonstrated in figure 15.3 in which one blue laser curtain panel passes in front of an adjacent panel.

Windows: Laser viewing windows are an option for laser barrier curtains but bear in mind that the window is typically a weak point in a laser safety system. A laser barrier is installed because of its laser damage threshold properties while windows are chosen according to optical density and similar specifications. Windows generally have much lower laser damage thresholds than the barriers into which they are installed.

15.6 Laser barriers—partition style

Selection of partition style laser barriers shares considerations and constraints with laser barrier curtains. In this section we will summarize the commonalities and elucidate the uniqueness.

Overview—laser barrier partitions

Laser barrier partitions, also referred to as laser screens, are comprised of a select barrier material installed to create a divider, or partition, between work areas, especially between laser and non-laser spaces. The typical construction, shown in figure 15.4, includes a rigid frame to which the barrier material is attached. Attachment may be achieved using rope or twine suspension with grommets in

Figure 15.3. Laser trap created with by-pass.

the barrier material. Barrier materials are also sometimes attached to a metal frame with hook-and-loop fasteners or mechanical fasteners.

Free-standing laser partitions are supported on cross-braces of heavy or reinforced metal that rest on the floor directly or to which wheels or casters are attached. The best systems feature cross-braces that swivel or detach to minimize space required for either shipping or storage. Floor level base plates with appropriate slots sometimes are used instead of cross-braces.

Cross-braces are essential but establish a deficiency: anything raised just above floor level that extends in a perpendicular fashion from a planar surface creates a trip hazard. Solutions to this are obvious and include (a) remove the wheels and allow the supports to rest directly on the floor, (b) paint the supports a bright or glow-in-the-dark color, (c) use only some of the supports (not recommended), or (d) use shorter length supports (assumes tipping hazards are not increased).

Laser barrier partitions may be delivered from the manufacturer fully-assembled with panels attached and wheel hardware in place. Some partitions will require one-time assembly with a couple of small tools. A third type are truly portable laser partitions which are designed for quick set up and take down, typically for field maintenance or mobile laser applications. We recommend that the portable types not be used for long-term protection—permanent problems should be solved with permanent solutions.

Laser barrier partitions are constructed from the same materials as laser barrier curtains described in the preceding section. Notable additions to the list of partition

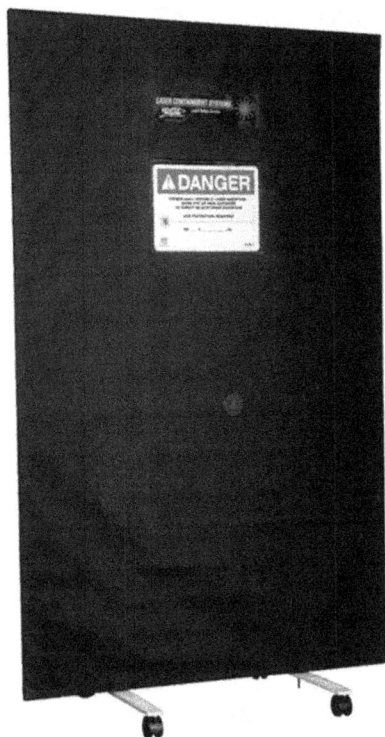

Figure 15.4. Typical laser barrier partition.

materials are composites and multi-layer honeycomb panels. Selection of the laser barrier materials, as with laser curtains, should follow laser damage and laser wavelength requirements plus the many non-laser considerations enumerated above.

15.7 Laser barrier partition design

There are a few additional considerations in the design of laser barrier partitions, both individually and when specified as part of a laser area isolation system. First, one must determine an appropriate height for the barrier materials. Most commercially available barriers are constructed to a height of just over 6 feet; this, somewhat arbitrary specification, evolved on the theory that eye level for most users is below this threshold and the risk of direct exposure to the beam is minimized. If there are indirect hazards, or raised and recessed floors, or a population of uniquely tall people, taller barriers should be considered.

Just as laser barrier partitions may vary in height, they also vary in widths. If you are designing a lab to include rigid barrier partitions, utilization of floor space will be optimal if actual barrier widths in total approximate the desired length of barrier runs. The weakest point in a wall created with multiple laser barrier partitions will be the joints or connections to the panels. Connector strips should have laser safety ratings at least as great as the barriers themselves. Connectors should attach firmly to both barriers as they are joined together and should leave no gaps between the connector strips and the barrier panels. Ideally, the connector strips, sometimes also called light blocker strips, will allow the barriers to rotate relative to one another and enable corners and turns in the partition wall.

A shortcoming in laser safety for partition systems mounted on wheels or casters is the gap created between the barrier and the floor. These gaps should be evaluated as potential hazards by the LSO. If no gaps at the floor are desired, skirts can be installed at the bottom of the laser partitions or the partitions can be dismounted from the wheels of the support systems and set on the floor.

A final point on laser barrier partition design concerns options. Laser safety equipment manufacturers have shown abundant creativity with the options that are available for partitions. Consider the following options as a partial list:

- *Viewing windows*: Laser viewing windows of mineral glass, acrylics, and other materials may be installed in laser barrier partitions. These windows should be selected to attenuate safely the laser radiation within the laser work area; the calculations and evaluation of the viewing hazard is similar to the process used in selecting laser goggles. An important and often over-looked caveat to the installation of a viewing window is that the window is typically the weakest point in the partition wall—the absolute laser damage threshold of a window is much lower than most laser barrier materials.

- *Doors and pass-thrus*: Egress for a laser area surrounded by laser barrier partitions is afforded by (1) establishing a vestibule or labyrinth using the partitions, (2) physically moving the barriers by breaking through a connection, or (3) constructing a doorway through one of the partitions. One variant is a hinged door cut into a metal laser barrier partition. The LSO should ensure that light traps are complete when such a door is in the closed position. A second variant is a fabric flap door that personnel may push through. A doorway through a partition wall affords opportunity to add interlocks, signage, entry and exit switches and other room entry control equipment. See the example metal laser barrier shown in figure 15.5.

- *Ceilings and roofs*: With some laser barrier partitions it is possible to construct a lightweight roof system supported entirely by the barriers. This is most commonly found in manufacturing environments where a temporary laser installation is needed in the middle of an open floor that has mezzanines or arcade offices on an upper level.

- *Other connections*: Connections to other equipment from the laser partitions is also possible. We have designed systems in which the movable barriers are integrated into one end of a laser barrier curtain or are attached to a wall or post using custom-built light blocking components.

15.8 Laser barriers for optical tables

Recognizing that creating a laser safe environment is often a journey and not an all-in-one event, especially for established laboratories, we offer as a starting point this basic tenet: your eye sightline should not enter into a direct path to an open beam source upon taking one or two steps into the interior of a laser laboratory. The

Figure 15.5. Laser barrier partition with door.

fastest, simplest way to achieve this one basic goal on an optical table with an open beam is to set a small metal laser barrier at the end of the table or even to surround the entire table in the appearance of an elevated hockey rink.

Short of building an enclosure for an entire system, laser safety for optical tables is enhanced by employing one of several types of smaller, modular barriers:

- Table surrounds enable creation of a low barrier wall, like a hockey rink, around all or part of an optical table. This type of barrier is shown in figure 15.6.
- End stops consisting of compact single panels attached at key positions on a table can block direct viewing of an open beam from the highest risk locations around the table. An example is a square metal sheet of sufficient dimension and rated for laser penetration installed at the end of a laboratory bench to make it difficult to have a direct line of sight from a laboratory entry to the laser source.
- A beam tube should be installed where there is risk of personnel putting hands and arms into the beam path. Ideally these are laser-rated metal tubes but other constructions may satisfy the LSO.

15.9 Laser barriers for windows

A final category of laser barriers is laser blocking products for use on windows. Typically the windows an LSO is concerned with are those that open onto parking

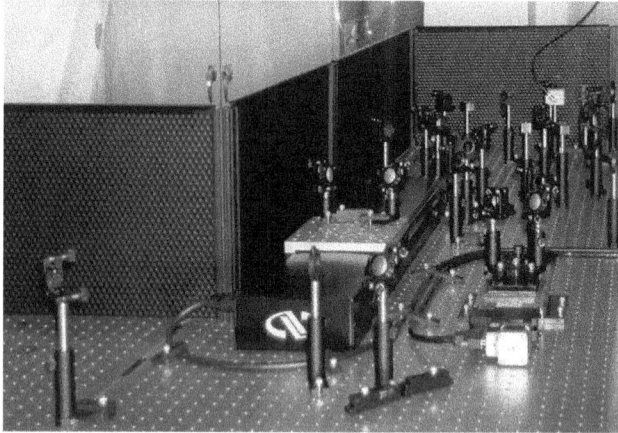

Figure 15.6. Optical table surrounds.

lots or face into adjoining spaces. Interior windows including windows on doors or observation windows from hallways may also require barrier materials. There are two basic forms of laser window barriers:

- *Laser window blocks*: A window block consists of laser blocking material, of either metal or fabric, that is shaped to the size for a particular window to which it is attached using fasteners such as hook-and-loop (like Velcro®), magnets, or other hardware. Accommodation for local storage of removable window blocks is highly recommended. For window blocks on doors in medical settings, we suggest attaching the window block below or next to the window when the block is not in use. The magnetically attached window block in figure 15.7 is easily affixed to the steel door when the block is not needed to screen the laser.
- *Laser window shades*: Roller shades, 'Roman blinds', and other forms of window shades of fabric laser barrier materials are all available. The best designs from a laser safety perspective minimize or completely eliminate light leakage through construction of side channels and window sill slots that encompass the laser barrier materials and create very tight light traps.

15.10 Laser barriers at a doorway

A common room configuration involving lasers, occurring especially in laboratories, features an open laser source in a room with a doorway to a public space such as a hallway or small office area. Depending on how the door or entry is configured, we may recommend the creation of a vestibule using appropriate laser barrier materials. This vestibule typically takes one of two forms: the closed vestibule which creates opportunities for tight control of the environment and the open vestibule which works best when external doors are where room control is centered.

Closed vestibule: A closed vestibule is constructed by laser barriers that must be moved or adjusted to enable access to a lab. For example, if the door is at a corner of the laboratory, a laser curtain may be suspended from a single straight track to

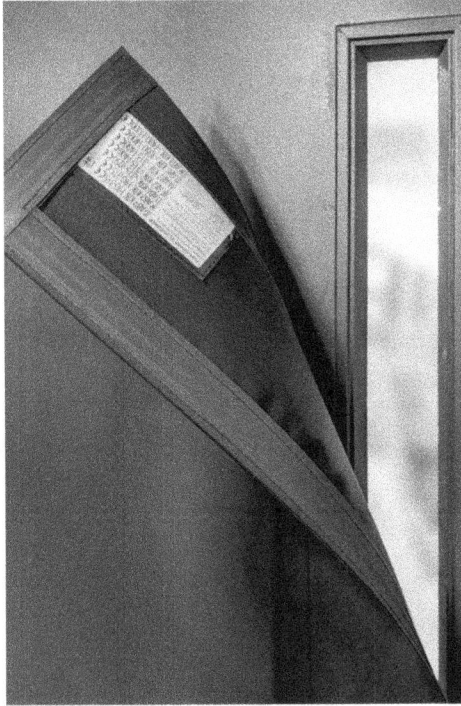

Figure 15.7. Window block with magnetic tape.

create a fabric 'wall' that must be opened to gain access into the working lab area. This fabric wall might be affixed to both walls and have a controlled opening in the middle, or it may be permanently attached to one wall and have a controlled opening where it meets the other wall. If the door is located more centrally along a wall in the lab, a vestibule can be created with a track layout for the laser curtain set in a semi-circle around the entry area on the inside of the lab.

The closed vestibule can be interlocked but may require other engineering and administrative controls. Any vestibule should have obvious ANSI Z136 laser warning signs posted at eye level on the visitor side of the laser barrier. Appropriate laser eyewear for the laser setup within the lab should be available to users within the vestibule with, ideally, some mechanism or procedure for identifying which laser filter should be used at the time of lab entry.

Figure 15.8 shows a shallow laboratory door vestibule created with a semi-circle of laser barrier curtain on a curved track. Entry to the laser controlled area is gained by moving the curtain aside by pushing through a break in the curtain wall. This type of setup may also be interlocked.

Open vestibule: An open vestibule is sometimes preferred, especially when door entry is controlled by a keypad or similar system. The open vestibule is typically created by suspending a laser barrier curtain in an 'L' shape at the door entry, thereby forming a maze or short labyrinth in the passage to the lab workspace. This configuration provides the lab entrant an opportunity to don appropriate laser safety

Figure 15.8. Laser room vestibule.

eyewear and/or time to assess the laser hazard beyond the curtain by conversing safely with staff on the other side of this laser barrier-induced beam trap.

15.11 Final thoughts

The laser safety standards in the United States, in Europe, and in other countries who have chosen to follow these guidelines or to create variants based on the established standards make it clear that personnel must be protected from laser radiation to the extent that is reasonably possible. So-called engineering standards in which a hazard is mitigated by design or by permanent or semi-permanent installation of fixed equipment will always be preferred over administrative controls and procedures for safety under which the potential for human error through uninformed decision or inattention to the environment is possible.

We hope that the notes we have presented in this chapter reveal that laser barriers can be designed and implemented to solve a range of laser safety problems. Laser barrier curtains can be hung to segregate laser areas, partitions can be set to create a wall, blocks can be put on windows, short barriers can be put directly on optical tables. Laser blocking materials are available in both fabric and metal representing a range of damage thresholds from which to select laser barriers. Read the standards, consult the experts, and design safety into your laser processing areas.

IOP Publishing

Laser Safety
Practical knowledge and solutions
Ken Barat

Chapter 16

US and European test methodology for laser protective eyewear

Michael D Thomas

16.1 Introduction

The proper evaluation of laser eyewear for optical density is an important topic which should be understood by both the manufacturer and end user of the product. A number of methodologies exist for the measurement and evaluation of laser eyewear as described in specifications put forth by both ANSI and DIN. These methodologies, the test procedures and the proper interpretation of the resultant data will be described in the following chapter. With either approach, the high optical densities found in today's modern eyewear, precipitated by the advances in laser technology and subsequent MPE, mostly preclude the use of spectrophotometers and require the use of lasers to properly evaluate blocking characteristics.

While both specifications call out measurement procedures for absorbing and reflective eyewear, in this chapter we will be focused primarily on the measurement of absorbers either through bulk properties of the material such as glass, or through the introduction of organic absorbers in a polymer matrix.

While these various test methodologies call out a full range of optical tests, all of which are germane for the proper evaluation, use, comfort and usability of laser eyewear, we will generally concentrate on the laser and laser test requirements put forth in the respective specifications.

16.2 ANSI and the Z 136.7 test specification

The ANSI Z 136.7 is the critical US specification for the measurement of laser protective eyewear. The specification provides a roadmap, outlines procedures and interpretation of test results for laser protective devices including eyewear and absorbing barriers. The specification outlines the measurement requirements, procedures, and methodology to perform laser transmission tests for absorbing material.

doi:10.1088/2053-2563/ab0f25ch16

As part of the ANSI procedures the laser power used to measure the samples is not defined. The optical density measurement can be made at any irradiance or fluence with the minimum detectability defined by the laser source and the response of the detector.

The specification does call out measurements for both CW sources and pulsed laser sources however.

16.2.1 Typical laser test setup

In an effort to describe the test requirements the following paragraphs will highlight some of the test and measurement requirements. This paragraph in no way attempts to lay out a full experimental configuration for the ANSI certification of laser eyewear, it is simply a basic overview in an attempt to describe the differences between the ANSI and the DIN requirement. For a more comprehensive review of the specification it is recommended that the specification be procured from ANSI and reviewed by the reader.

The test setup defined by the ANSI specification simply calls out a laser source at appropriate wavelength and a detector to measure the incident optical power and the transmitted optical power. Quite often this requires a detector which is linear over a detection range of many orders of magnitude. Given the paucity of sensitive detectors at wavelengths beyond the range of silicon and InGaAs this measurement can be quite difficult for high optical densities.

In the pulsed case either an energy meter, pyroelectric detector or fast semiconductor detector can be used to determine the transmitted light through the sample. In the CW case an absorbing power meter, or a calibrated semiconductor power meter can be used.

16.2.2 Laser saturation

As described above, the ANSI specification provides specific insight into the proper measurement techniques and methodology for saturation measurements. In addition, there are requirements for knowledge of the average power, spot size, polarization, pulse energy, and spatial profile and temporal profile. The specification does not call out that any specific parameters be used in order to perform the tests however.

One requirement though is the measurement of a pulse's sample over a specific pulse width range in an effort to determine if saturation may take place in the absorber.

16.3 DIN and the EN 207 test specification

The EN 207 is a specification which has been adopted by the European Union as a general test set for laser protective eyewear. The specification deals with many aspects of the laser eyewear pertaining to the use including optical properties, VLT and ballistic protection.

The principal difference between the ANSI and the European EN 207 laser eyewear specification is the idea of a *resistance class* or resistance level to laser radiation, and a wavelength/pulse width range over which that resistance class is defined.

The resistance class requires that for a certain optical density, defined by the MPE of the working environment, the protective eyewear must be able to withstand a specific exposure level to either a certain irradiance in W m^{-2} or fluence in J m^{-2} without the optical density of the protective device dropping below the requirement. This can happen through saturation, material damage, or worst case, penetration.

In addition, the specification provides a minimum exposure for each level of protective optical density, wavelength range, and laser pulse width. This approach to defining specifications for laser eyewear is significantly more complicated that that required under the ANSI Z 136.7 and the following verbiage served to better define and help the user understand the requirements, test conditions and end use of the specification.

16.3.1 Wavelength range considerations

Different wavelength ranges interact with protective materials in significantly distinct ways. In some cases the host material acts as a bulk absorber, limiting transmission at wavelength ranges without the introduction of organic dyes or absorbing materials into the host matrix. The EN 207 specification accounts for these differences by dividing the wavelength ranges into three different categories.

16.3.1.1 Wavelength range 1: 180 nm–315 nm
The most popular lasers in this wavelength range include excimer, and 4th and 5th harmonics of Nd:YAG based laser systems. Because wavelengths shorter than 180 nm do not propagate through the air, the short wavelength cutoff is set at 180 nm. Special precautions, however, may need to be taken for deeper UV wavelengths.

Many glasses have high absorption within this wavelength range and standard undoped polycarbonate also exhibits high absorption within this range of laser wavelengths.

16.3.1.2 Wavelength range 2: >315 nm–1400 nm
This wavelength range comprises the bulk of the lasers used today for research, medical applications, material processing, ranging, and designation. The wavelength range comprises the UV through the visible to the near IR and includes many common lasers both pulsed and CW.

16.3.1.3 Wavelength range 3: >315 nm–1000 μm
This wavelength range begins at the near infrared and extends out into the far infrared. Sources within this range include thulium, and holmium lasers as well as CO_2 systems. Lasers within these wavelength ranges operate mostly pulsed or CW.

16.3.2 Pulse width considerations

In the DIN EN 207 specification each wavelength range is broken down into four pulse width windows. These include D which corresponds to CW irradiation, I which is Q-switched, R which is considered long pulse and M which includes pulses shorter than 1 ns.

Each of the pulse width ranges will have a different effect on the ability of the eyewear to both provide the correct optical density and address the specific resistance class.

16.3.2.1 CW irradiation (D)

Using CW lasers to test for optical density and resistance class poses the most difficult challenge to penetration and damage to laser eyewear tested under the DIN EN 207. Because of the infinite duty cycle, at resistance classes commensurate with modest optical densities thermal absorption and subsequent temperature rise can cause cracking and melting in glass or glassy materials and burning, melting and penetration in polycarbonate or plastic absorbing eyewear. Reflective laser eyewear which only absorbs a small fraction of the incident laser power may be better suited in applications which require a high optical density and subsequent high resistance class.

In applications where true CW lasers are not available to test to the D requirement the specification allows for the use of high repetition rate Q-switched or mode locked lasers with appropriate scaling as shown in the requirement.

16.3.2.2 Long pulse lasers (I)

The I class of lasers covers pulse widths from 10^{-6} to 0.25 s. Lasers of this type can be usually found in medical or material processing systems. Lasers which may operate within this pulse width range include Er:YAG, Er:GSGG, Ho:YAG, or Nd:YAG.

16.3.2.3 Q-switched lasers (R)

Class R lasers include Q-switched laser system which operate with a pulse width between 10^{-9} and 10^{-6} s. This would include scientific Nd:YAG lasers and other visible, UV and infrared sources which include either a Q-switch or fast electrical discharge to generate a 'giant' laser pulse.

Q-switched lasers present a unique problem in measurements of resistance class and optical density because of the high peak irradiance of the laser pulse. This high peak irradiance may in some cases saturate absorbing laser dyes and above a certain threshold either limit the ability of a dye to absorb based on the concentration of dye in the plastic matrix or in a worse case saturate the laser dye to a point where the optical density is reduced below the MPE required by that specific eyewear.

16.3.2.4 Ultrafast lasers (M)

The M designation refers to any laser pulse width which is shorter than 1 ns. With the advent of ultrafast laser systems this nomenclature could support pulse widths in the 10 fs range to the 1 ns range. This comprises over six orders of magnitude.

Because of the large pulse width range appended by this level and the significant changes to material properties (including absorption) when irradiated by ultrashort laser pulses, it is important that the user or LSO be especially wary of eyewear marked for the M specification.

Because of the nature of ultrafast lasers the bandwidth of the laser source tends to be wide. In some cases the eyewear may provide significant protection at the peak wavelength of interest but may not offer protection or may not offer sufficient protection at wavelengths away from the peak.

In addition to the wide bandwidth, ultrafast bleaching has been observed in both organic dyes and in certain types of glass used for laser protective eyewear.

16.3.3 Laser parameters used in DIN EN 207

Independent of both the resistance class, wavelength range and pulse width range are certain laser parameters which must be used to be in compliance with the DIN EN 207. These parameters include spot size, repetition rate (PRF), number of pulses (exposure duration).

16.3.3.1 Laser beam profile and spot size considerations

When performing measurements to the DIN EN 207 specification a beam with a Gaussian spatial profile must be used to make the measurement. The Gaussian profile assures that there are no 'hot spots' within the beam and that measurement parameters of the beam are standardized. The requirement calls out a 1 mm (1/e) beam diameter for D, I and R lasers and a 500 μm (1/e) beam diameter for M type lasers.

If a beam with a diameter larger than 1 mm is used for the measurement the specification contains a scale factor which scales the irradiance or fluence defined by a particular resistance class.

16.4 Performance of EN 207 specification

16.4.1 Significance of the read across

Resistance class testing must take place within one of the specific wavelength ranges called out in the EN 207 specification in order to certify a laser eyewear sample. Quite often, however, a laser is operational at a specific laser wavelength within that operating range and there may be a desire to use the protective eyewear over multiple wavelengths within that range. An example of such usage would be a PPE which protects against both the main Nd:YAG laser wavelength at 1064 nm and the second harmonic at 532 nm. Another example would be in the case of a diode pumped YAG laser where protection would be needed for both the pump diode wavelength and the primary laser wavelength.

In this case as long as the resistance class is measured along with the optical density at the peak absorption wavelength a *read across* may be made using a spectrophotometer. In all cases the maximum resistance class which may be marked is the measured resistance class at the primary wavelength.

In many cases a spectrophotometer cannot provide sufficient signal/noise to measure the optical density of high OD samples. In this case the high OD samples can be pressed or made thinner to be adequately measured, the subsequent OD calculated from the thinner sample.

16.5 Conclusion

It can be seen that there is a significant difference between the data generated using the ANSI Z 136.7 and the DIN EN 207 measurement techniques. The concept of a *resistance class* gives the LSO an understanding for the level at which damage, and a subsequent compromise to the laser eyewear may occur. The DIN measurement also allows a comparison between different manufactures' products because many of the test parameters are bound within specific ranges of fluence, irradiance, PRF and pulse width.

Chapter 17

Elements and considerations in designing and/or selecting a room interlock system

Patrick Bong

17.1 Introduction

Interlock systems are intended to warn personnel of the presence of hazardous laser radiation and inhibit laser operation when the conditions of the interlock are violated. The control measures for each of the four laser classifications will differ based on the hazard to be mitigated.

17.2 Hazard assessment

Interlock systems may be applied to Class 1 and Class 2 laser systems; however, they are primarily applied to Class 3B and Class 4 lasers. Interlocks should be applied after an assessment of the hazards and the probability of personnel exposure to hazardous laser radiation. To determine the applicability of interlocks to a laser system the hazard assessment should be performed in three steps.

1. Consider the mechanisms that could lead to the exposure of unprotected personnel.
2. Determine if the considered exposure event is probable.
3. Apply substitutions or engineering controls to eliminate the hazard.

As an example, an unmitigated light path through an open door could be a hazard to persons on the other side of the door. The addition of interlocks to the door can inhibit laser radiation when the door is open.

17.3 Minimal system for attended operation

A minimal interlock system may be used when a laser is intended to operate only when a laser operator is present. This interlock would include a warning light to alert personnel that there is laser radiation inside the enclosure, and an Emergency Shutoff button to inhibit the laser. It is recommended that a switch on the door is used to inhibit the laser when the door is opened. The door switch will prevent the operation of the laser in the absence of a laser operator.

Emergency shutoff buttons, also called E-Stops, must be a distinctive pushbutton switch or mushroom-type pushbutton. A sign near the button should indicate that the function of the button is for 'Emergency Laser Shutoff'. It is recommended that the button is red with a yellow background, and when activated the button should latch. To return to operation the actuator must be manually unlatched by pulling, twisting, or released by a key. Key release type actuators should be used when the Emergency Shutoff requires a reset by a system manager.

The interlock circuit should be connected to the external interlock input on the laser power supply. The interlock circuit may also close a laser shutter in addition to the external interlock or when an external interlock input is not available.

The circuit in figure 17.1 shows a basic interlock circuit with a relay coil that latches, across relay contact K1-1, when the reset button is pressed. It is recommended that the circuit is powered by a 24 V power supply for electrical safety. The reset button should be a momentary contact that is open when the button is not being pressed. The circuit is faulted when the Emergency Shutoff button is pressed or the door is opened. The Laser On warning lamp is lit, and the laser power supply external interlock is closed when the circuit is reset and the relay is energized.

24 V illuminated laser safety signs are available as commercial off-the-shelf (COTS) items from many laser safety vendors. The optional Laser Off warning light is used to indicate that there is no laser hazard when using a sign that can indicate both the Laser On and Laser Off states.

Additional relay contacts may be used to enable laser shutter controllers or drive laser shutters directly. Capacitor and resistor are usually added to the power supply side of a relay contact when a contact is used to operate a laser shutter solenoid. The capacitor provides additional current to the solenoid when the shutter is opening and the resistor reduces the current to the 'hold open' value necessary to keep the shutter

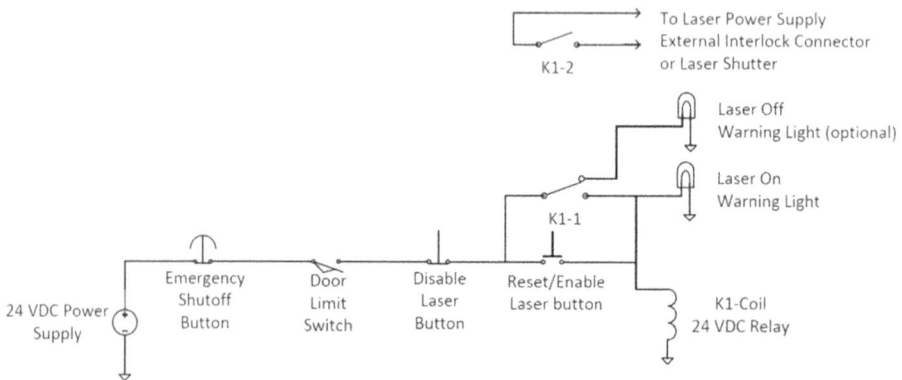

Figure 17.1. A basic interlock circuit with a relay coil that latches, across relay contact K1-1, when the reset button is pressed.

open and reduce the load on the power supply. The resistor and capacitor values are usually available from the manufacturer of the laser safety shutter.

17.4 Interlocks for unattended operation

A more elaborate interlock system is required when lasers are operated in the absence of a laser operator. The interlock system is similar to the one used for attended operation, however, the door switches are not optional. The door interlocks ensure that laser radiation is inhibited when the area is accessed by unauthorized personnel. When this level of system is used the enclosure is called a controlled laser area.

Door interlock switches must detect that access doors are closed. Doors that do not provide access to the controlled laser area may not require an interlock switch. Be advised that a door that only provides emergency egress should be interlocked to inhibit laser radiation in the event of an emergency.

Doors around the perimeter of a controlled laser area may allow the door interlock to be bypassed temporarily to allow laser operators to enter or exit the area while the laser is operating. The door interlock bypass is typically 15 seconds in duration, enough time to allow several people through a door. Often a bypass may be re-triggered to allow for a longer bypass period to allow equipment or tools to be brought into or out-of the controlled laser area.

Doors that are not normal access points to the controlled laser area should be designated as non-defeatable doors. Non-defeatable doors may not be bypassed at any time and will inhibit laser radiation when opened.

Access to a controlled laser area is typically provided by an access keypad. The access keypad provides the signal to bypass the door interlock. The door may then be unlocked by a user key, or a magnetic door lock may release the door when the correct code is entered in the access keypad. Alternatively, the access device may be a key switch, to prevent the sharing of keypad codes, or an electronic card reader.

An illuminated laser safety warning sign at each access point to the nominal hazard zone should indicate the No Hazard Laser Off and the Laser On Warning. The positive indication of either state provides a clear and unambiguous notice to personnel entering the area of the state of laser radiation in the area. A sign has malfunctioned when neither indicator is lit, and personnel should access the area as if laser radiation is present.

An Arming Station is provided inside the controlled laser area to prepare the area for laser radiation. The Arming Station may be either a key switch or a button that switches the illuminated laser safety warning sign from No Hazard Laser Off to Danger Laser On (your wording may differ due to regional requirements). Laser shutters and power supplies may be enabled after the controlled laser area is armed. When a laser system consists of several power supplies and shutters (such as pumped power systems) individual elements may be controlled by Hazard Permit Stations. Hazard Permit Stations can also be used to control groups of elements that require control as a set.

A minimum of one Emergency Shutoff button should be located inside the controlled laser area. Additional Emergency Shutoff buttons should be located inside the controlled laser area when rapid access to an E-Stop is impeded by laser tables, structural elements, or other mechanical barriers. The placement of an Emergency Shutoff button outside the controlled laser area should be considered, to allow emergency personnel to inhibit laser radiation, when there is a possibility of First Responders accessing the area.

17.5 Reach back cascade

Some laser systems present an extreme hazard when the laser radiation is not abated after an interlock is faulted. This may occur due to a laser shutter that is stuck in the open position or is blocked by an obstacle. The shutter can be monitored by a switch that senses that the shutter is in the closed positon. If the switch does not close within the time period required for the shutter to reach the closed position, then the shutter is presumed open and additional measures are taken to render the area safe. One such method is called a reach back cascade. The reach back cascade uses a timer to monitor the door interlocks. If the area is violated then the timer starts. If the shutter switch does not detect that the shutter is closed before the timer ends, then the laser power supply external interlock is opened and the power supply is turned off.

17.6 Design considerations for interlock systems

Individuals responsible for developing, implementing, and maintaining interlock systems should have the appropriate education, training and skills to perform these responsibilities.

Systems must be electrically safe and conform to electrical standards and codes.

System components should be protected from damage and conspicuously labeled to reduce the likelihood of inadvertent modification.

Chapter 18

Paperwork considerations (not documented not done)

Ken Barat

18.1 Introduction

No safety program goes without paperwork, in today's world maybe it is better to say no program exists without documentation. The old saying not documented not done, is as true as when it first was spoken, regardless of whether the documented is in ink or electrons.

In laser safety several programmatic elements need some type of documentation. Some are more obvious than others. How much paperwork one wishes to generate will have a lasting effect on one's program and maybe even safety compliance. Both LSOs and user can be buried by too much paperwork, to the point it becomes more of a paperwork exercise than serving any useful purpose.

So, a balance needs to be met. Of course, if there is an incident, there never seems to be enough documentation to prove how hard you tried and how innocent you are. So, I will let you choose which of these documents you feel you need to have in your own laser safety program. A number of these documents are expanded on in this chapter.

18.2 ISO 9001, just a quick word

There is an international standard being used by many firms worldwide, ISO 9001, which lays out several quality management concepts and steps. An interesting topic that falls under the general heading of paperwork, is the difference between 'records' and 'document control', which is addressed in ISO 9001.

Records are an important organizational asset; they provide the primary route for evidence-based verification and traceability and are able to demonstrate compliance with regulatory and or customer requirements.

Documents are considered as specifications or procedures and their supporting medium (e.g. paper or electronic). The standard implies that over time these documents will evolve as new information supersedes old and that change must be managed. Hence documents are active and dynamic.

Records, on the other hand, are looked upon as being static since they are historical in nature. They are the documents that state the results of activities undertaken in accordance with the product realization, measurement, analysis and improvement processes (e.g. calibration logs and non-conformance or corrective action reports). They also provide evidence that an activity was performed in the manner specified (e.g. inspection records).

18.3 Record retention

An item many LSOs overlook is their own institution's rules on retaining records. Most people take it for granted that medical records need to be kept for a long period of time, but are either in the dark or less clear on records they may be generating or reviewing. How long should training tests be kept, interlock logs, old copies of SOPs? The answer to how long, or if at all, falls within the policies of the institution.

Clause 4.2.3 ISO 9001-2008 states that an organization must control the documentation required by the quality management system and that a suitable document control procedure must be implemented to define the controls needed to approve, review, update, identify changes, identify revision status and provide access. The document control procedure must clearly define the scope, purpose, method and responsibilities required to implement these parameters.

So, from the list in table 18.1 it is safe to see that, as in any safety program, a level of documentation is required. A few words will be said about common errors in each of the documents listed. There are three that are the most important to demonstrating a working laser safety program.

Let's highlight some of these 'document friends'.

18.4 Training records

There have been several laser accidents where no record could be found that the individual involved had received any laser safety training. Sometimes, it might have been that training at the facility was never given, or people were counting on previous training. Sometimes, just sloppy record keeping, from failing to sign an attendance sheet to the record being lost. The most common errors on training records are: not recording the data, not having access, date errors.

18.5 Standard operating procedures

The ANSI Z136 laser standards and IEC 60825 standards all call for a SOP for Class 3B or Class 4 laser use. The SOP is a hazard evaluation document. The SOP outlines what are the hazards involved in the laser work, as well as non-beam hazards and how they will be mitigated. The most common errors with SOPs are: failure to address all hazards, failure of users to sign the document, failure to update as changes are made that affect safety, and failure to renew in a timely manner.

Table 18.1. Members of team paperwork, coming your way by your friends at Bureaucracy.

1. EHS laser safety chapter
2. Training record (institutional basic laser safety course)
3. On the job training record
4. Training record for non-beam hazards
5. Risk assessment
6. Standard operating procedures (SOP), initial sign off
7. Renewal of SOP
8. User signatures indicating reading of SOP
9. Refresher training record
10. Room interlock checks
11. Alignment eyewear approval
12. Temporary control area document
13. Alignment procedures
14. LSO qualifications
15. Required optical density for eyewear and windows
16. Testing of barriers
17. Use area audits
18. Action on corrective actions from audits or regulatory inspections
19. Accuracy of warning and contact signs and postings

Good practice and common-sense say that all users working under a particular SOP need to read that SOP and sign off that they have. If the SOP is calling out hazards and how to mitigate them, how can one say they understand the controls if they have not read the SOP? A signature is validation that the document has been read.

18.6 Audit records

The ANSI Z136 standard requires evaluation of control measures and how effective they are on a periodic basis. So, what are the most common errors? Not performing the survey in a timely basis, failure to document corrective actions, unclear on who is responsible for corrective actions and follow up on corrective actions. Audits need to reflect the work being reviewed, we may need to tailor some audits to the work setting, one size may not fit all.

18.7 Laser safety chapter

A laser safety chapter, some place where your institutions polices are laid, out is critical. This is not so much your program goal, which is the same everywhere to make sure no one gets hurt, but rather what one must do to be an approved laser user. When we say laser user we mean Class 3B and Class 4 lasers or laser systems. In today's world if one searches the web for laser safety chapter or policy, there will be no end to the number of chapters one can take material from or to receive guidance from. In fact, if you look at enough chapters you will start to notice about 3–8 models that are being used over and over. A laser safety chapter, once developed, should be reviewed at least every two years. There may be a new use or policy one may like to add. One example of that is saying something about the use of laser pointers or any restrictions on them for use on your site.

18.8 Accuracy of warning and contact signs and postings

Why have a warning sign on the entry to a laser use area? It is a hazard communication device. Therefore, to be useful it needs to be accurate. Just saying Class 4 laser in use, tells one they need to take care, but really does not impart much useful information. That is why the sign should indicate the wavelength and optical density of any eyewear that might be required. Who to contact in an emergency or problem, if not accurate is useless and wastes precious time as well as reflecting poorly on one's program.

18.9 Alignment eyewear approval

The use of alignment eyewear can be a great aid to the laser users. But let's be clear, alignment eyewear does not provide the same level of protection as full protection eyewear. Which seems obvious from their names. Alignment eyewear allows enough transmission of visible wavelengths to be seen by the user, while allowing laser radiation from a direct hit at a Class 3R level and one's aversion response to make a user move their head before damage can be done.

From the perspective of the LSO one is allowing eyewear that will not provide full protection. If one reads the laser standards the role of laser protective eyewear is to reduce any laser transmission to or below the MPE; meaning, if one uses alignment eyewear a possible transmission above the MPE is possible. Therefore, some institutions believe the LSO should give specific permission for the use of alignment eyewear. One of the goals of giving permission is to see if the need for alignment eyewear is real, i.e. 'can alignment be done another way?' as well as reviewing the eyewear selection.

18.10 Temporary authorization/temporary work authorization

This is a document that LSOs and users can find very useful. When? Well, how about when equipment is purchased and a check of its specifications is needed, open air alignment that will be enclosed, such as at some user facilities with beam lines. Another time such a document or authorization can be useful is on short-term project use and during service of Class 1 systems or products. The form below demonstrates the elements of such a temporary authorization.

Temporary Work Authorization form

Work as described below may be performed during the stated period after all required concurrences and authorizations have been obtained.

Effective date: _____   **Expiration Date**: _____

Work Location: Building _____   Room: _____

Maximum duration: two weeks

This document will be posted during the duration of its application at the work location

Work Scope (describe work including permitted and prohibited activities, boundaries and "stop points" as appropriate):

Controls required I am issuing TWA as Laser Safety Officer per ANSI Z136.1 & ANSI Z136.8. My hazard evaluation conclusion is that engineering and temporary administrative controls listed below are sufficient to mitigate the laser hazard.

1.

2.

3.

4.

5.

Note: At least one temporary physical barrier must be specified to control access. If it is infeasible to implement a physical barrier a justification must be provided.

Personnel included in this authorization (signature denotes verification that training in the provisions of this Temporary Work Authorization has been provided)

Work Leader _____

                Name   Signature   Date

(Work Leader/supervisor is responsible for assuring that all required training, including job- and task-specific training, is provided prior to beginning work)

Users:

_____
Name   Signature   Date

_____
Name   Signature   Date

_____
Name   Signature   Date

_____
Name   Signature   Date

Concurrences and Work Authorization

Principal Investigator Concurrence _____

                          Name   Signature   Date

EH&S Concurrence _____

                (Print Name of LSO) Signature   Date

18.11 Interlock checks

If one is taking safety credit for access door interlocks, those interlocks should be checked to confirm their functionality. Many users will tell you the interlock works because it crashes at least once a week. This is not the same as a functional check. The operational check needs to include all elements of the interlock systems, shutter(s), lights, emergency stops, door sensors, etc. If a problem is encountered, documentation is required of the corrective action.

18.12 Conclusion

A long list of records have been presented with only a small number of follow-up details presented. If any of the remaining ones apply to you it is your responsibility to clearly understand what they entail.

Chapter 19

Explaining engineering control measures found in standards

Ken Barat

19.1 Introduction

Regardless of whether you fall under ANSI standards or IEC standards, you will find that engineering and administrative control measures come into your laser work. Some are mandated to follow while others are suggested, a 'should' rather than a 'shall'. Often users have difficulty either applying these control measures or understanding them. The goal of this chapter is to explain the engineering control measures, some of which trace back to the earliest days of laser standards and laser technology.

On the application of control measures, users and management need to remember that the standards give the LSO a great deal of latitude in their application. The LSO can develop alternate controls. In addition, the ANSI standard does not support the use of redundant control measures, once the system is made safe.

Administrative controls, training, procedures, and eyewear are all covered in other chapters, so let us take a closer look at engineering controls.

19.2 Engineering controls

Many engineering controls can be traced back to product safety engineering controls, when early standard writing committees felt the need to support such controls.

Below are the common engineering controls found in most laser standards.

19.2.1 Protective housing

You will find wording such as 'A protective housing shall be provided for all classes of lasers'. Which of course does not explain what a protective housing is. So one now goes to the definition section for clarification. The common definition is 'An enclosure that surrounds the laser and prevents access to laser radiation'. If you read on into the definition it gets to the protective housing preventing access to electrical component hazards and may enclose optical elements. A protective housing has a slightly different definition than an enclosure, which of course just clouds what the term means. Take the line from ANSI Z136.1 section 4.4.2.1 'If a user-created enclosure does not meet the requirements of a protective housing (e.g. a non-interlocked cover), it shall be considered as a barrier or curtain'. Which just seems to muddy the water to a new level. *So*, before we get too lost, the cover over the optical cavity (optical resonator) is the protective housing, which is why all class lasers have one.

19.2.2 Interlocks on removable PH

For personal safety and equipment safety all protective housings should be interlocked. The issue with laser standards is just about every time the word 'all' is used there is an exception given. The same goes for protective housing interlocks. For the standards allow interlocks to be replaced with a warning label, indicating the housing is not interlocked, which at times is useful to laser system manufacturers and even users.

19.2.3 Service access panel

This is a panel which is intended to be removed, which allows access to laser radiation associated with a Class 3B or Class 4 system. A panel or door that is open to put samples or parts into the system, would not fit this definition. The service access panel needs to be interlocked or be opened with a tool and be properly labeled. The tool necessary for panel removal should not be one's fingers, so wingnuts are frowned on. The tool option, while it seems less safe than an interlock, is actually very useful to system designers.

19.2.4 Key control

When I read how the laser standards describe 'Key Control and Master Switch' it is really an on and off button. Therefore, I am not sure it deserves to be called out as a separate engineering control.

The master switch shall produce beam termination or system shut off and shall be operated by a key or code access. A single master switch on a main control panel is okay for multiple laser installations where the operational controls have been integrated.

For key control, the key cannot be removed when in the on position and the laser or laser system cannot be operated unless the key control is in the on position.

Traditionally the key control has also served as an access control tool. In an earlier version of laser standards, the key was required to be removed when the system is not in use. In many ways key control is comparable to key control found in x-ray machines.

19.2.5 Collecting optics

Some laser systems use collecting optics, lenses, telescopes, microscopes, and eye loupes. The control for collecting optics states a means shall be incorporated, such as interlocks or filters, to maintain the laser radiation transmitted through the collecting optics to levels at or below MPE. In addition, permanently mounted collecting optics housings containing filters sold other than as an integral part of the product shall be labeled with optical density and wavelength for which protection is provided.

Cutting through standards language, if one looks through an optic into a Class 4 or Class 3B system it needs to protect the viewer. As to labeling separate viewing items such as a fiber viewer, many are not labeled with what protection they provide.

19.2.6 Area warning device

This relates to Class 3B and in particular Class 4 use areas. The control states an area warning device that is visible prior to entering the control area is needed. The stated purpose is to ensure that persons who are about to enter the area are aware that a laser is emitting or is about to begin emitting accessible laser radiation within the area.

The key to me is the term *accessible laser radiation*. So many area warning devices are illuminated lights that come on as soon as power is applied to the laser system. Which adds to the problem of people not paying attention to warning signs, for often when the light is on there is no hazard.

It is interesting that the standards describe a 'Visible Warning Device' as one that indicates when the laser is operating. In the use of visible warning devices, the illuminated sign is the common approach. So when one's visible warning device is a single indicator, the actual safety or risk potential is not clearly indicated.

What challenges such signage is when there is more than one laser system or multiple lasers not slaved to the same pump laser. So just having a single light does not indicate which laser is on or is producing accessible beams. One of the best answers to this is a flat screen display, maybe tied to a programmable logic controller system (PLC) which can show the status of all lasers in the control area. While this approach sounds expensive and complex, it is only a matter of time before its value and use drives cost down.

Of course, another and less expensive option is to have the sign tied to a shutter. When the shutter is down, but the laser powering up, the sign is off. When the shutter is up allowing beam propagation the sign is on. There are issues, such as can the shutter be moved and if so how does one remember to replace it? But these items are all solvable.

Multi-status signs are another option, but once again, having more than one system in the laser control area is a challenge.

19.2.7 Laser radiation emission warning (usually visible)

This item is one of the most watered down controls, it should be improved or just retired. It starts out as an indication to those in the laser control area that the laser is emitting or getting ready to emit laser radiation. It can be achieved by an audible signal or a light. Then the standards retreat to saying a light on the laser or control panel is sufficient. The emission indicator on a laser is usually a small LED, which can only be seen if one is looking for it or standing near the laser. The same applies if it is on a control panel, one has to be at the panel.

Few facilities use a verbal countdown or alarm to indicate the laser is about to emit its beam. All of this should be handled by communications between the users in the laser control area regardless of whether the system is Class 3B or Class 4. Once it is on and work starts it becomes invisible to the users.

If the laser control area is made up of several isolated areas and set ups, then an indication of which areas have beams becomes of greater critical importance.

It goes without saying, but I will say it anyway, any visible indicator must be able to be seen with one's laser protective eyewear on.

19.2.8 Emergency conditions

The standards say: For emergency conditions there shall be a clearly marked 'Emergency Stop' or other appropriate marked device suitable to deactivate the laser or reduce the output to or below the MPE.

This is certainly a legacy item, dating from when all laser systems required a great deal of electrical power or related hazardous systems. In today's laser world, unless related to robotic work, I am unsure how useful or cost effective this requirement is. This is why some places consider that the on/off switch on the power supply or a labeled circuit breaker achieve this requirement. The use of the red mushroom button is fading. Also, an item of miscommunication is the use of an off-the-shelf Emergency Stop (E-Stop) button. A first responder might get the impression that pressing it will remove power from the laser, while it might just be dropping a shutter and the laser is still energized.

19.2.9 Class 4 entryway controls

All the laser standards accept that Class 4 laser use environments present a potential hazard. With that said, then access to this area needs to have some safeguards or controls. On the flipside, the standards recognize laser use areas can have a great deal of variation. Therefore, the LSO needs to evaluate the proper level of access control. With that in mind the LSO has three options.

19.2.10 Non-defeatable approach

What this means is entry in or out will cause the laser system to default into a safe mode, power off or a shutter to block the beam. This approach is rarely used, due to its all-or-nothing approach.

19.2.11 Defeatable controls

Meaning the trained and authorized individual has a means to bypass the interrupt mode and enter without disturbing operational mode. Usually achieved with an interlock system that has a card key, key pad or key system. The majority of these also incorporate a system where, if the entry door stays open past a preset time limit, the system falls to a safe mode. This time limit is generally acknowledged to be from 15 to 30 s. Defeatable is the most common approach one finds in Class 4 use areas.

19.2.12 Administrative control

If the LSO deems that an interlock entry way system is not feasible, other means of access control are allowed to be used. This can be as simple as a warning sign, such as found in operating rooms to an electronic lock (which has no interface with lasers), which can keep those without the proper code or swipe card from entering.

Just using a key lock, has the problem of too many people having 'master Keys'. With the use of administrative controls additional training might be required.

19.3 Conclusion

As stated at the start of this chapter, many engineering controls found in laser standards are also laser product safety requirements. In the research setting many homemade lasers do not have all the required product safety items, which is acceptable for one-of-a-kind systems. It is when a system is designed for commercial use or technology transfer that the product engineering controls become relevant. One of the most misunderstood controls concerns interlocked access control. Just remember, just because it is a Class 4 laser system does not mean an access interlock is required.

Chapter 20

Dye laser, hazards and good practice for safe use

Ken Barat

20.1 Introduction

The dye laser systems has had a bumpy history. While on the surface it offers the capability to have multiple wavelengths from one system, that is limited to one wavelength at a time, or one dye system at a time. Its biggest drawback is that the dye and solvents tend to be hazardous to people. Therefore, it requires special handling and disposal. But if one is looking for a source of various visible wavelengths the dye laser system might be your best option. This chapter will give a general overview of dyes, handling and related issues.

20.2 Dyes and solutions

Dye lasers normally use a lasing medium composed of a complex fluorescent organic dye dissolved in an organic solvent. Animal experimentation has shown these dyes to vary greatly in toxicity and potential carcinogenicity; consequently, all dyes should be treated as hazardous chemicals. A few, especially DCM (4-dicyanomethal-2-methyl-6-pdiethylaminostyryl-4-H-pyran) have been found to be very strong mutagens. In many instances, the solvent in which the dye is dissolved plays a major role in the solution's hazards. Practically all solvents suitable for dye solutions are flammable and toxic by inhalation or skin absorption. Some dye solutions come pre-mixed from the manufacturer, in which case efforts must be made to determine which dye and solvent were used for the preparation. It is important to read the safety data sheet (SDS) for each of the dyes and solvents used in the laboratory.

20.3 Preparation for dye work

Any person/employee working with dye laser systems and components needs to receive safety training, in particular, chemical safety training. Any class given need not only contain general chemical safety training but a module on dye lasers and the solvents involved. The importance of reading the relevant safety data sheets cannot be over emphasized. Since the majority of the chemical components are flammable, storage of such chemicals also needs to be reviewed and understood. You need to know the location of any eyewash stations or showers.

20.4 Supervisor and staff responsibilities

To re-emphasize, the group supervisor and staff should prepare by doing the following:

- Identify which laser dyes and solvents will be used and review safety data sheets.
- Perform an assessment of the hazards and controls that will need to be carried out to reduce any possible exposure.
- Make sure the hazard and controls have been documented in any operating or assessment document.
- Ensure everyone handling the materials has taken the required chemical safety training.
- Posting needs to reflect the correct chemical hazard.

20.4.1 Equipment, concerns

Has a location for mixing and handling the dyes been assigned? A fume hood is the best location when mixing laser dyes or when handling them if that handling may generate an airborne hazard, i.e. vapors, mists or fumes.

Good practice requires spill pans under pumps and reservoirs, on occasion some labs will enclose the set up. Tie down tubing at either end to make sure it does not become disconnected.

20.5 Real work rules

- Do not eat, drink, smoke, chew gum, apply cosmetics or store food or beverages, in work areas where laser dyes and solvents are being used.
- Use mechanical pipetting aids when handling dye solutions.
- Keep containers of solvents and dye solutions closed.
- Cap off and/or drain dye lines that are not in use.
- Keep the work area clean. Use wet methods for housekeeping in dye work areas. Remove visible stains as much as practical during cleanup.
  Note: Housekeeping should not be expected to clean up dye spills.
- Keep flammable solvents in approved storage cabinets.
- Wash hands after handling laser dyes and solutions.
- Personnel who have had skin, eye, or inhalation exposure to dye powders or solutions should get medical assistance.
- Minimize the quantity of pure dye or solutions.
- Caution placards at entrances to work areas.

Dye laser trick: run tubing through a pail of sand to reduce pump vibrations. *DO NOT pour dye down the DRAIN!!!!!!*

20.5.1 Protect yourself

- Wear safety glasses with side shields.
  These can be replaced by a face shield.
- Wear a laboratory coat.
  This can be replaced with a suitable apron.
- Closed-toe shoes will be worn when handling these materials.
- Wear the proper protective gloves.
  Note: Some solvents like DMSO will be absorbed through the skin and carry along other chemicals.
- Use of a respirator is rare.

20.6 If there is a spill

Depending on your institution you maybe required to notify the safety division for them to either perform the clean up or assess what is needed, such as respirator use. If you must clean up yourself, treat as if you were mixing the dye. Wear gloves, dispose of any wipes or towels as hazardous waste. Try not to spread the spill and make the contaminated zone bigger.

20.7 More on dyes

For most dyes, little is known about their toxic properties, except that they are often members of chemical families that contain highly toxic materials. Lawrence Livermore National Laboratory, a US Department of Energy facility, is one of the few facilities I am aware of that has done an extensive review of common laser dyes.

Below is their assessment, please note this data was determined in the 1970s.
Key
L = Limited control class; M = Moderate control class; S = Strict control class.

20.8 Laser dye/solvent control classes

Material	Synonym	Control class	Comments
BBQ	M	Nonmutagenic	Unknown toxicity
Benzyl alcohol	L	Moderate toxicity	Low vapor pressure
Carbazine 720	M	Nonmutagenic	Unknown toxicity
Coumarin 1/460	M	Nonmutagenic	Moderately toxicity
Coumarin 2/45	M	Nonmutagenic	Unknown toxicity
Coumarin 30/515	S	Mutagenic	Unknown toxicity
Coumarin 102/480	S	Strong mutagen	Unknown toxicity
Coumarin 120/440	M	Nonmutagenic	Unknown toxicity
Coumarin 314/504	M	Nonmutagenic	Unknown toxicity
Coumarin 420	M	Nonmutagenic	Unknown toxicity

(Continued)

Coumarin 481	M	Nonmutagenic	Unknown toxicity
Coumarin 498	M	Unknown mutagenicity	Unknown toxicity
Coumarin 500	S	Mutagenic	Unknown toxicity
Coumarin 540A	M	Nonmutagenic	Unknown toxicity
Cresyl violet 670	S	Very strong mutagen	Unknown toxicity
1,3,5,7-Cyclooctatetrene (COT)	M	Unknown mutagenicity	Unknown toxicity
DCM	S	Very strong mutagen	Unknown toxicity
p,p'-Diaminoquaterphenyl	S	Mutagenic	Unknown toxicity
p,p'-Diaminoterphenyl	S	Mutagenic	Unknown toxicity
Dioxane	M	Moderate toxicity	
DMSO	M	Moderate toxicity	
DODCI	M	Unknown mutagenicity	Unknown toxicity
DQOCI	M	Unknown mutagenicity	Unknown toxicity
DPS	M	Doubtful bacterial mutagen	Unknown toxicity
Ethylene dichloride	M	Suspected carcinogen- avoid inhalation	
Ethyl alcohol	L	Low toxicity	
Ethylene glycol	M	Moderate toxicity	Low vapor pressure
Fluorescein 548	M	Unknown mutagenicity	Unknown toxicity
Kiton Red 620	L	Nonmutagenic	Practically non-toxic
Kodax Q-Switch #2	M	Unknown mutagenicity	Unknown toxicity
Kodax Q-Switch #5	M	Unknown mutagenicity	Unknown toxicity
Nile Blue 690	S	Commercial grade is strongly mutagenic, purified dye is not	Unknown toxicity
Oxazine 720	M	Nonmutagenic	Unknown toxicity
Rhodamine 4	M	Nonmutagenic	Eye irritant, slightly toxic
Rhodamine 6	M	Nonmutagenic	Eye irritant, slightly toxic
Rhodamine 6G/590	M	Special case	
Rhodamine 110/560	S	Weak mutagen	Unknown toxicity
Rhodamine 610/B	M	Nonmutagenic	Moderately toxic
Rhodamine 640	M	Nonmutagenic	Unknown toxicity
Sulforhodamine 640	M	Unknown mutagenicity	Unknown toxicity

Dye pump in secondary containment, photography plastic trays, metal would be better (figures 20.1 and 20.2).

Figure 20.1. Dye pump in plastic trays.

Figure 20.2. Dye stains on tile floor.

Chapter 21

Laser disposal, end of life cycle thoughts, hospice for your laser

Ken Barat

21.1 Introduction

As laser technology advances, lasers become outdated or applications change such that particular lasers are no longer needed, especially in universities and research institutions. Typically, these lasers go to storage; but, eventually they go to salvage. For disposal personnel unfamiliar with the individual components of lasers, they may be disposed of in routine trash. However, laser components, which vary by laser, require careful consideration when it comes to disposal.

21.2 Why should you care?

Ceramic collars once broken can release beryllium, dye lasers can contain carcinogenic materials, excimer lasers may contain fluorine and e-waste items, just a few of the reasons why one should be concerned with how old lasers are disposed of.

In addition, if one decides to sell or give away old lasers it is critical to make sure whoever receives it knows how to use it safely. Do you want someone hurt or hurting someone else with a laser that can be traced back to your institution?

21.3 E-waste

There is general public awareness of the issue of e-waste. Government organizations have generated regulations and guidance to deal with the disposal of electronic products. There are products that fall under the radar, with a number of laser systems falling into this grouping.

Your laser safety program or hazardous waste department should consider how to deal with lasers at the end of their life cycle or waste stream. One idea is that equipment can only be donated to facilities that have an individual responsible for laser safety, a laser safety officer (LSO), to assure of safe use of the laser systems. Otherwise, to prevent the possible misuse of Class 3B and Class 4 systems, they are to be made inoperable and properly disposed of. One can also choose to go with trade-ins or sale to resellers. Resellers must be aware of export restrictions on technology.

At this time, two groups have a responsibility to see that the laser systems are handled in the proper fashion, the user (present owner) and the institution's property management salvage department.

21.4 What are my responsibilities?

Different persons may have contact with the laser as it goes to the disposal process. A series of questions and action points should be targeted toward each group. This is essential because the steps needed predisposal are usually unclear within an institution. For example, within an organization, there are people responsible for the transportation of waste and hazardous materials; however, their knowledge of the laser equipment is usually limited.

21.5 Questions for the user

(1) Have you contacted property management to ask which forms need to be filled out?

(2) If you are disposing of a dye or excimer laser, have you flushed out the chemicals in the pump containers, tubing and inner cavity? For instructions, please see the dye laser or excimer laser sections of this chapter.

(3) Do you have the user manual for the laser? If so, send it for disposal with the laser.

(4) Did you contact your LSO to see if they can find a new home for your laser?

(5) Remember to remove the laser from your inventory record.

21.6 User responsibilities

When a laser is ready to be released for surplus, the user should find the laser system user manual and send it along with the laser. While commercial lasers should have several labels on them, the most important for surplus are the manufacturer label and the logo label. From the manufacturer label, one obtains the model and serial numbers. The logo label contains wavelength and output data and might indicate the general laser name (i.e. argon, HeNe). Before sending the laser to surplus, the user should place a sign on the laser indicating its optical laser medium (i.e. Nd:YAG, dye, argon etc). This will be a great help to the property disposal group especially if the user manual cannot be found.

21.7 Questions for hazardous waste transporter or handler

(1) If it is a dye or excimer laser, ensure the user has flushed out the chemicals in the pump containers, tubing and inner cavity and disposed of them correctly or saved them for you.

(2) The majority of power supplies built prior to 2000 with capacitors will contain capacitor oil, see that the oil is drained and placed in the proper container. Laser will have a label indicating date of manufacture.

(3) Any other type of laser has no special precautions required for transportation.

21.8 Questions for surplus receiver

(1) Have you received the proper paperwork?

(2) Is the laser type identified, so the proper disposal steps can be taken?

21.9 General approaches to laser disposal

There are several alternatives to laser disposal which may be considered. Some laser manufactures have a cradle-to-grave service and will accept old lasers for recycling as a service to the user. Surplus personnel are advised to contact the laser manufacture and ask if the laser system may be returned for disposal, refurbishment, usable components, or possible equivalent to new repair. However, not all manufacturers have this option, and some require the user to pay for shipping.

It is also possible to transfer the laser as a donation to another university's engineering or physics department etc. It is necessary to ensure that the laser system complies with all applicable safety instructions for operation and maintenance and that the receiving program has a viable laser safety program. A 'Limit of Liability' document will also need to be generated. If it is a 3B or 4 laser or laser system that will be transferred, the Laser Safety Officer (LSO) will need to be contacted. They may contact the designated institution to see if they have an LSO and the steps that will be taken to use the laser in a safe manner.

If it is determined that neither of those options are viable, the laser will be disposed of. The following is an excerpt on general laser disposal:

Lasers, should not to go to public auction, that is the opinion of this author. This is due to concerns over the misuse of the laser system; one example is lasers being used to expose commercial aircraft pilots while in flight. The laser must be rendered unusable.

Knowing the manufacturer and model number, a call to the manufacturer is prudent to check on any possible hazardous material components.

21.10 Power supplies

The simple action is to cut off the plug and as much as possible of the AC source cord. However, lasers that utilize electricity as their main source of energy and manufactured prior to July 1, 2006 have lead in the printed circuit boards. Dispose of them as electronic waste. Also consider that many laser systems utilize a high voltage capacitor system hence an electrical shock hazard is a real possibility. Standard capacitor safety needs to be followed, use of ground hooks, removal of capacitor oil, etc.

21.11 Optics

For optics, consider removing the optics and place them in a ziplock bag. Label the bag with the laser manufacturer information and then send them to the LSO, others might need them.

Various lasers require different actions, below are examples found in this chapter.

21.12 Dye lasers

Active concerns

These lasers use a liquid medium. This medium is composed of organic dyes and solvents, all which must be considered carcinogenic or mutagenic. Dye is transported by means of plastic tubes.

Disposal

Wear personal protective equipment. This includes goggles, particle masks, chemical resistant apron, and gloves. Rinse pump containers, tubing and inner cavity several times with methanol, then with water until internal circulating fluid appears to be clear. All washing materials must be considered hazardous waste and dealt with as such.

Common dyes and solvents include coumarin, rhodamine, exalite, stilbene, oxazine, and DMSO and Rhodamine 590 Tetrafluoroborate.

21.13 Excimer lasers

Active concerns

Excimer lasers use a combination of halogen and noble gas. Each has its own risks depending on the quantity and concentration. It also includes an internal electrical system that is required to make the two families of gases form a dimer.

Common excimer lasers include argon fluoride, hydrogen fluoride, krypton fluoride, and xenon fluoride.

Disposal

User should have flushed out the resonator prior to sending to surplus as well as any pre-mix chambers. Wear gloves and safety glasses. Remove chamber and crack open in a well-vented area. Remove any circuit boards and dispose of as electronic waste. Remove power cord. Hazardous materials to consider include argon, fluoride, krypton, and xenon.

21.14 Diode/semiconductor lasers

Active concerns

From a size perspective 95% of these laser units are made up of heat sink, current controls etc. The actual laser diode is smaller than a paper clip. Some diode laser systems may be part of a diode fiber system.

Common laser diodes include gallium aluminum arsenide, gallium arsenide, indium gallium aluminum phosphide laser.

Disposal

For individual diode units or arrays, simply break the housing unit, taking care to wear protective eyewear. When multiple units are received at one time they should be treated as e-waste. For fiber optic systems, while wearing safety glasses and protective eyewear, the fiber should be cut near the diode end and fiber segment put in a sharps container. One of the hazardous materials to consider is gallium arsenide (hazardous waste).

21.15 Diode/telecommunications laser systems

Active concerns

In a majority of cases, these lasers contain optical fibers, which are used to deliver laser radiation. Most, if not all, injuries come from handling the fibers. Treat them as potentially 'sharp' components.

Common telecommunication lasers include gallium aluminum arsenide, gallium arsenide, indium gallium aluminum phosphide laser, and ytterbium-doped fiber.

Disposal

Fibers are to be treated as sharp optical components and need to go into sharp, disposable container. Safety glasses and gloves are to be worn when handling (cut or remove from diode on laser box). Diode is to be treated as e-waste, as above. One of the hazardous materials to consider is silicon.

Part IV

Problems and solutions, are you
dealing with these?

Chapter 22

How are you dealing with these topics?

Ken Barat

22.1 Chapter note

This section deals with several common problems found in many a laser lab. When a commercial product is mentioned by name, it is because the author has seen it in use and is not familiar with other company names, sort of like how Kleenex has become a universal term for facial tissue or Xerox for copying. There is no compensation to the author for mentioning these firms. First let's start by listing some good practices:

22.2 20 smart work practices—all are important, number sequence does not relate to safety or order of importance

1. Doors need an access control approach to prevent injury to unauthorized personnel.
2. The proper warning signs should be posted.
3. The illumination in the area should be as bright as practicable in order to constrict the eye pupils of users.
4. Where practical, the laser system or beam should be enclosed to prevent accidental exposure to the beam.
5. The potential for specular reflections should be minimized by shields and by removal of all unnecessary shiny surfaces.
6. Check for stray reflections, often and after optics manipulation.
7. Remember 750 nm–820 nm beams can be seen as faint reflections, but that does not indicate true power or irradiance of beam.
8. Windows to hallways or other outside areas must be provided with adequate shades or covers when necessary to keep the nominal hazard zone (NHZ) within the room.
9. The main beams and reflected beams should be terminated or dumped. This is required for any accessible laser for which the MPE limit could be exceeded.

10. Electrical installation must meet electrical safety standards.
11. The active laser should never be left unattended unless it is a part of the controlled environment.
12. Warning devices must be installed for lasers with invisible beams to warn of operation.
13. The laser work area should be maintained as free of clutter as possible, to minimize the chance of accidentally igniting something.
14. Ensuring that lasers are well secured to the work surface helps prevent a stray beam.
15. Never look directly at the laser beam.
16. Clear all personnel from the anticipated path of the beam.
17. Before operating the laser, warn all personnel and visitors of the potential hazard, and ensure all safety measures are satisfied.
18. Be very cautious around lasers that operate at frequencies not visible to the human eye.
19. Do not wear hanging ID badges, reflective jewelry (watches) or other objects.
    a. Wedding rings can be covered with tape, i.e. painters' paper tape.
20. Use proper eye protection when working with a Class 3B or Class 4 laser.

22.3 Ventilation

It is not uncommon for a laser user to have heartache over the ventilation system in their room. This is not the same as temperature control, which is another heartache. Rather it is the concern over down drafts and particulates coming down from the air in and out system. Many times the air in vent is located right over a piece of sensitive equipment that down drafts disturb.

One possible solution is a product called an 'Air Sock'. This is a fabric system that uses diffusion to facilitate the air exchange in the area (figure 22.1). It comes in varying sizes and configurations. It allows air exchange and traps any particulates inside the fabric system.

22.4 Access control

For a Class 4 laser, user area access control is required and desired. The most often heard remark about interlocked rooms is 'I know the interlock works, it crashes my system at least once a day'.

This is not a happy refrain. If one does not have beams leaving the optical table a successful solution is an electronic lock that is not connected to the laser (figure 22.2), neither its power supply nor shutter. These systems just control access. If one really wants to spend money such systems can be expanded to tie into one's training database, blocking any who do not have the required training or whose training has lapsed from entering. All of these need some mechanism for a first responder to enter, maybe a special key, card or combination.

Figure 22.1. Air sock example.

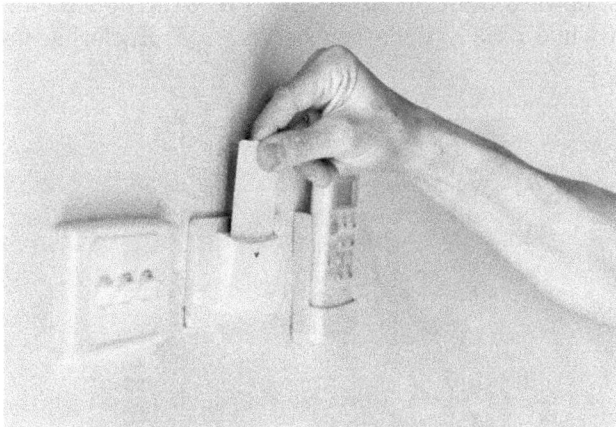

Figure 22.2. Electronic lock examples.

22.5 Housekeeping/storage

Housekeeping in a lab is harder than keeping one's room straight. There always seems to be more stuff than space. Where is the TARDIS technology when you need it? One of the best ways to obtain good housekeeping habits is to have a set clean-up day. Whether it is once a week, or more likely once a month, when everyone knows when clean-up day is, schedules and activities can be planned. My experience has been the more this gets done the shorter time it takes. After a while people get into the habit of putting things away and this scheduled day goes from a full day to a half or shorter.

In addition, there are products one can buy for the storage of items. Many of these are designed for workshops and garages, check out those catalogs for ideas, i.e. ULINE (figures 22.3–22.5).

22.6 Layout of optics

The layout of optics and related items in a research setting is an art. Today two successful approaches take advantage of CAD technology. This has gone from an extremely expensive software and luxury to a free downloadable software or at least reasonably priced. Not only is CAD drawing great for setting up one's optical table but also to see how equipment will fit within your workspace.

Another approach is if a large size plotter/printer is accessible, print off a life size plot of your set-up. Then lay over one's optical table (figure 22.6).

22.7 Periscope

The rotating polarizer and periscope optics are considered the two most dangerous optics on an optical table. They have either been involved in numerous stray beam incidents or have a great potential to get the user in trouble. Focusing on the periscope optics, upward beams need to have a block to contain a stray reflection. The beam path while usually open can be enclosed. Enclosing these beams will

Figure 22.3. Storage examples for optics etc.

Figure 22.4. Poor housekeeping.

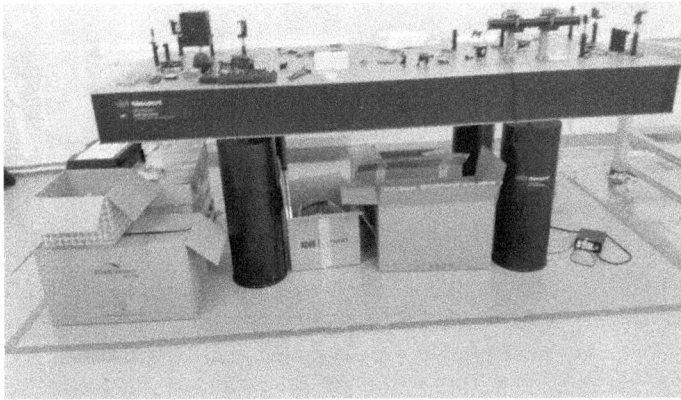

Figure 22.5. Fire hazard under table.

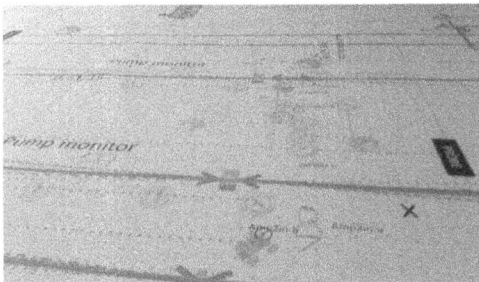

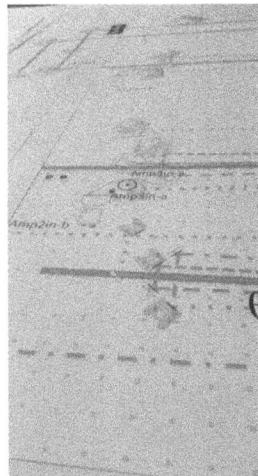

Figure 22.6. Large plotter examples.

stop fingers and tools from passing through them. Here are a number of solutions (figure 22.7) By the way, paper is never a good beam block, but is often used for that purpose (figure 22.8).

22.8 Cable and hoses

Wire and hoses are usually on the floor or in trays (figure 22.9). On the floor they are trip hazards (figure 22.10). Sometimes the floor can look like a swamp of snakes

Figure 22.7. Periscope beam control examples.

Figure 22.8. Block for periscope beams.

Figure 22.9. Cables in floor.

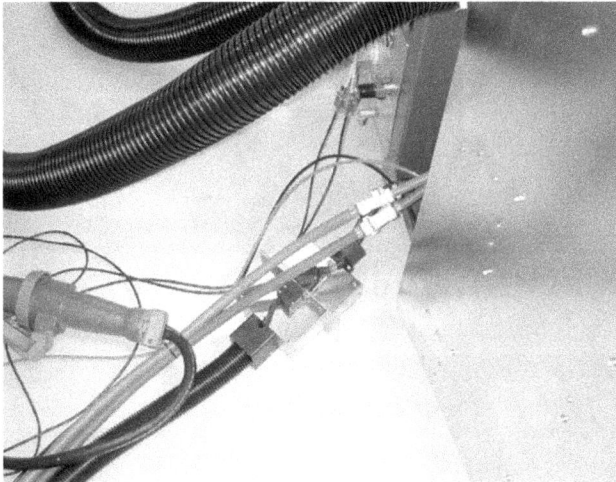

Figure 22.10. Cooling lines over electrical line.

(figure 22.11). Cable trays can be found in different widths and can be connected. Commonly they are made from a plastic material, some will even hold up to being run over by a forklift. This does allow running of wires and fluid in common trays but separate slots.

The other option is the false floor, rare in a routine laser lab but has been used in clean rooms and large laser areas such as petawatt set ups.

Leak sensors, while a maturing technology that even offers wireless systems that will send out a message to one's phone, from my experience have found limited use in research laser labs. But this is certainly worth consideration.

22.9 Optics and cable identification

This seems rather simple, but it is not a common practice (i.e. putting a label on optics and components that are part of one's optical set-up) (figures 22.12–22.14). If you

Figure 22.11. Cable and hose covers.

Figure 22.12. Labeling of components.

remove a lens or mirror from a post, it may be difficult to tell its purpose. This also extends to cables as well. Some folks have found this practice useful and a time saver, others see no value in it, which camp are you in?

22.10 Pump diffuse scatter

So many laser systems use a green pump light, but how the diffuse scatter from the pump to the next laser system component is handled varies a great deal. A first good

Figure 22.13. Labeling of Pockel cell.

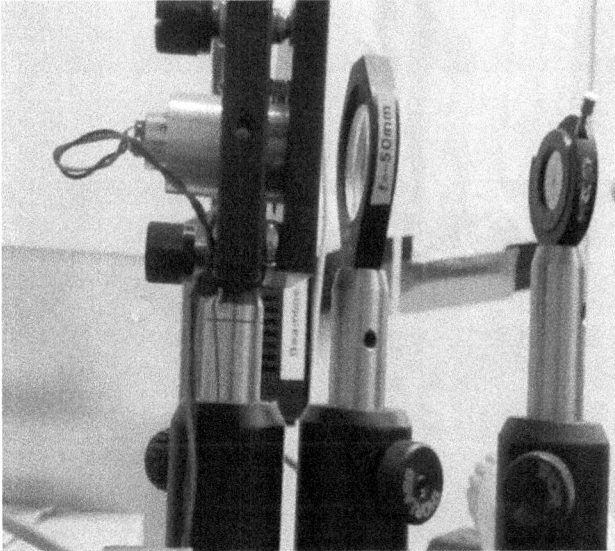

Figure 22.14. Labeling of optical mount.

step is to contain the short open distance in a tube. Many stop at this step. The result is diffuse green scatter (figure 22.15). Which is below an eye injury level but can not only light up the table but cause eye strain. The next step up is to have the beam tube opaque to contain the scattered light (figures 22.16–22.18). This can be done in several ways, from wrapping the tube to using an opaque metal or plastic tube. Many times there is a small gap, this too can be blocked.

The bottom line is that no matter how one achieves it, the diffuse green scatter should be contained.

Figure 22.15. Diffuse green pump light.

Figure 22.16. Optics covers.

22.11 Signage on entryway door

Laser warning signs on one's door, while a simple concept, can seem to be all over the place (figure 22.19). Is it one sign, combining all laser information, is it one sign for each laser, what about hazard communication signs for other hazards, or contact information? It seems like every institution has its own idea on this. The ANSI

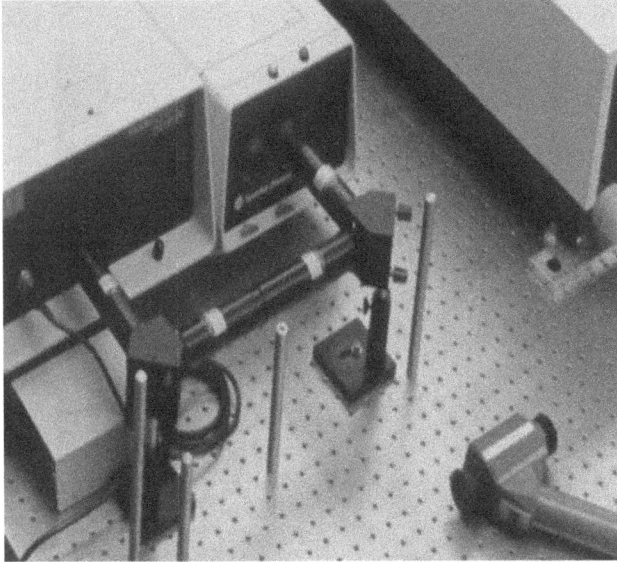

Figure 22.17. Beam tube application.

Figure 22.18. Foam used to block diffuse scatter.

standard requires a warning sign at the boundary of the hazard zone. It allows information on several lasers to be on the sign (really no limit). The research standard Z136.8 says that you can just say multiple lasers in use, check with operator for proper eyewear (figure 22.20).

Figure 22.19. Too many laser warning signs on door.

Figure 22.20. No information about eyewear on sign.

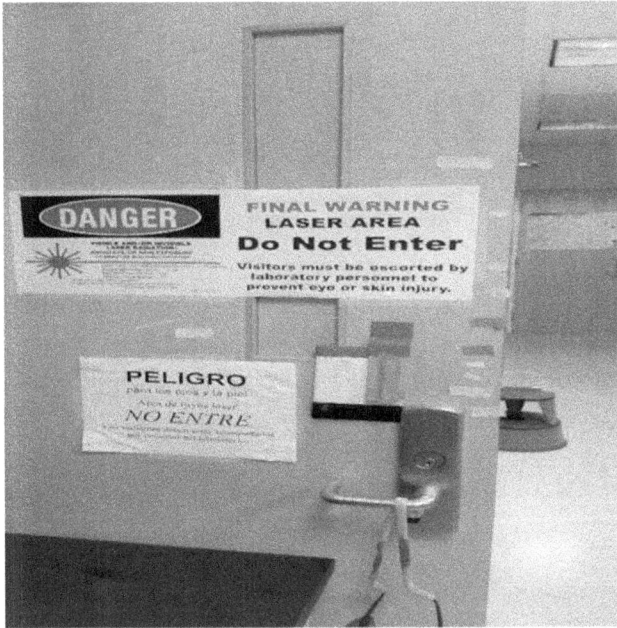

Figure 22.21. Door left open, but check out warning.

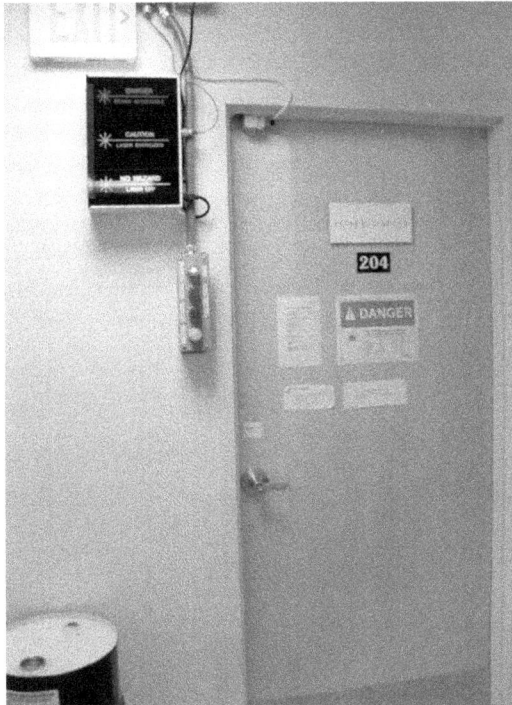

Figure 22.22. Interlocked entry door.

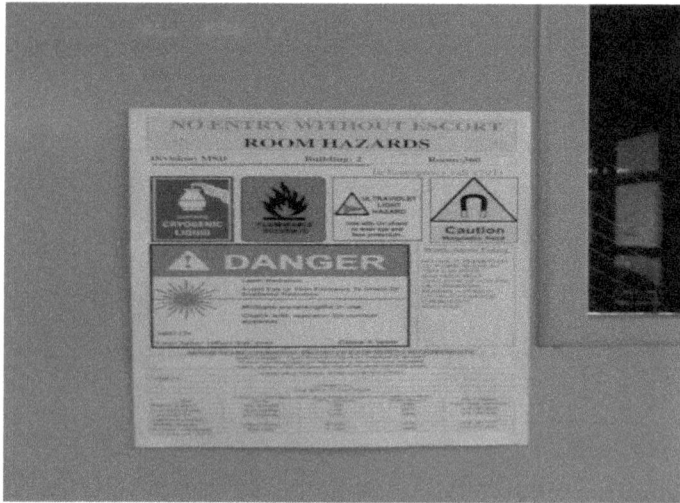

Figure 22.23. Combination sign, reduces clutter on door.

While at Lawerence Berkeley Nat Lab, my students and I developed an all-in-one laser warning sign and hazard communication sign that reduced door clutter and met the requirements of the chemical safety officer. Figures 22.21–22.23 show some examples of how laser lab doors look today, you decide on which way you want your door to look.

References AKA additional reading

[1] ANSI Z136.1 20014 *Safe Use of Lasers*

[2] ANSI Z136.7 *American National Standard for Testing and Labeling of Laser Protective Equipment*

[3] ANSI Z136.8 *American National Standard for Safe Use of Lasers in Research, Development, or Testing IEC 60825-1, Ed 2.0 (2014), Equipment classification and requirements*

[4] ANSI Z535.2-1998 *American National Standard Criteria for Safety Symbols*

[5] ANSI Z535.3-1998 *American National Specification for Accident Prevention Signs*

[6] IEC 60825-1, Ed 2.0 (2014) *Equipment classification and requirements*

[7] IEC 60825-14, (2014) *Safety of laser products—users guide*

[8] OSHA *Occupational Safety and Health Standards for General Industry, Code of Federal Regulations, 29 CFR 1910*

[9] Barat K 2013 *Laser Safety in the Lab* (Bellingham, WA: SPIE Press)

[10] Barat K 2015 *Laser Lab Design, SPIE Spot Light Series, e-book*

[11] Barat K 2008 *Laser Safety: Tools and Training* (Boca Raton, FL: CRC Press)

[12] Barat K 2006 *Laser Safety Management* (Boca Raton, FL: CRC Press) ch PPE

[13] Barat K 2019 *Understanding Laser Accidents* (Boca Raton, FL: CRC Press)

www.ingramcontent.com/pod-product-compliance
Lightning Source LLC
Chambersburg PA
CBHW080539220326
41599CB00032B/6311